Learning and Categorization in Modular Neural Networks

Jacob M.J. Murre

MRC Applied Psychology Unit, Cambridge

Routledge
Taylor & Francis Group

LONDON AND NEW YORK

First published 1992 by
Lawrence Erlbaum Associates

Published 2022 by Routledge
4 Park Square, Milton Park, Abingdon, Oxon OX14 4RN
605 Third Avenue, New York, NY 10017

Routledge is an imprint of the Taylor & Francis Group, an informa business

Typeset in 10/12 pt Times
by Mathematical Composition Setters Ltd, Salisbury, Wiltshire

Library of Congress Cataloging-in-Publication Data

A CIP catalog record is available from the publisher

ISBN 13: 978-0-805-81337-1 (hbk)
ISBN 13: 978-0-805-81338-8 (pbk)

To my parents

To My parents

Contents

Preface and Acknowledgements

This work describes a new algorithm for learning neural networks called CALM, from Categorizing And Learning Module. CALM is a module able to categorize and learn an arbitrary input pattern, either with or without supervision. The work is divided into three parts. Part I introduces the basic module and the learning algorithm. Part II describes its application to psychological modelling, pattern recognition and network design by genetic algorithms. Part III consists of an evaluation, with emphasis on the biological, computational and psychological plausibility of CALM models.

In a number of appendices, the implementation of CALM and similar modular neural networks in parallel hardware and software is discussed. These appendices include an analysis of performance on transputer networks (Appendix B2), a report on a realized 400-processor neurocomputer (Appendix B3), an outline of an implementation in analog hardware (Appendix B4), and a discussion of three software packages developed for the simulation of modular neural networks (Appendix B5).

The main theme in all chapters of this work is *modularity*. In almost all instances we will consider the characteristics of modularly constructed neural networks. Most of the research reported concerns the basic CALM module and its learning algorithm. Two principles are important in this learning algorithm. Categorization and learning are controlled by a *noise-driven search process, that is structured through competitive selection.* As will be argued, this principle is fundamental to widely different processes of self-organization in nature, ranging from the evolution of species and ecosystems to the recognition of a flying insect by a frog, or the settling of spins in a magnetic field. Another important principle is *locality*. This says that all processing is governed only by locally available information. In contrast to common practice in neural networks, this excludes the use of hidden variables and normalizing factors in the formal description of the algorithm. We adhere to the principle of locality in order to approximate the biological reality more closely and to facilitate implementation in parallel hardware.

The general principles are complemented by several more specific principles. A pivotal concept, derived from psychology, in the CALM algorithm is *arousal*. The 'arousal level' in a module is determined by the relative novelty of an input pattern. It is implemented as a self-induced control of learning and categorization. We suggest that the principle of self-induced arousal may be responsible for the dissociation between explicit and implicit memory tasks. According to Graf and Mandler (1984; Mandler, 1980), tasks involving explicit memory rely on conscious recollection and on reference to the context of the situation to be remembered. Implicit memory is said to be present when previous exposition causes a measurable improvement on some task, even if the subject is not consciously aware of this. Dissociation between the two types of memory is revealed in experiments with amnesic patients, with anaesthetized subjects, and with normal subjects. Graf and Mandler (1984; Mandler, 1980) have introduced two types of learning to explain the dissociation. In the CALM algorithm, we have linked these types of learning to 'arousal' and to the distinction between old and new input patterns by the system.

The CALM module derives its basic internal structure from biology. The wiring of the CALM module is loosely based on the neural structure of the neocortical minicolumn. Inhibitory connections are mostly short range, long-range connections are mostly excitatory. A single node (formal neuron) is either excitatory or inhibitory, but not both (Dale's law). Processes in nodes only rely on information that is locally available through synapses or through locally dispersed neurotransmitters (principle of locality).

Although primarily developed as a psychological model, the CALM algorithm may also be used for practical applications. Part of the motivation for the implementation efforts described in the Appendices is based on the conviction that, if the CALM algorithm is implemented in sufficiently fast hardware, it may be applied to practical tasks.[*] Because CALM networks are modular, implementation in parallel hardware is greatly facilitated. Moreover, the modular architecture can be shown to increase several aspects of the performance of neural networks, such as learning time, interference and generalization. Self-induced arousal may contribute further towards increased performance.

This book gives an account of research performed at the Leiden Connectionist Group of the Unit of Experimental and Theoretical Psychology at Leiden University. I would like to thank all members of the Leiden Connectionist Group for helpful discussions and for their pleasant company during the past 4 years. The work on the CALM algorithm is the result of joint efforts with Hans Phaf and Gezinus Wolters. I want to thank both of them for this period of fruitful cooperation. I also wish to thank Patrick Hudson for initiating and sustaining much of the work of the Leiden Connectionist Group, in particular concerning the hardware implementation of neural networks. I owe thanks to the following persons for their support, whose comments have assisted me in improving this work: Jaap van den Herik, Gerard Kempen,

---

[*] For the same reason, we have applied for a patent (see Murre et al., 1990a). This limits the application of CALM in commercial projects, but it leaves open the possibility to use CALM in scientific research projects.

Doug Mewhort, Willem-Albert Wagenaar, Gerard Zieleman and Jan Zieleman.

Part of the work reported here has been published elsewhere. This has been indicated either in the text itself, or in footnotes. I wish to thank my co-authors for their cooperation in providing me with material for this work: Bart Happel, Jan Heemskerk, Jaap Hoekstra, Patrick Hudson, Leon Kemna, Steven Kleynenberg, Hans Phaf and Gezinus Wolters. I also wish to thank Eric Postma for letting me reprint his figures of the ELAN model. Many students have participated in the various projects of the Leiden Connectionist Group reported in this work. I wish to acknowledge the help of Bert Boekema, Theo van den Bout, Rob Hartsuiker, Frans Kleer, Cor Kroon, Arend Melissant, Mirko Pelgrom, Freek Smouter and Wouter Trienekens. Their specific contributions have been cited in the text at the appropriate places. Without the cooperation of the above-mentioned students and researchers the writing of this book would have been impossible. I thank them all.

J.M.J. Murre
Leiden, February 1992

PART I

CALM:
Categorizing And Learning
Module

Chapter 1

Introduction

1.1 The importance of learning

The study of learning systems is important, both from a theoretical and from a practical point of view. Theoretically, the study of the ability of information processing systems to learn may result in better models for understanding how living organisms adapt to environmental demands by selecting and storing information. From a practical point of view, it may solve some programming problems encountered in the development of complex, intelligent information processing systems. Because a learning system may show autonomous programming, it can develop its own structure and function in interaction with the environment, without the need for detailed instructions by a human controller.

The recent interest in connectionism (Feldman, 1981; Feldman and Ballard, 1982), parallel distributed processing (Rumelhart and McClelland, 1986; McClelland and Rumelhart, 1986a), neural networks (e.g. Hinton and Anderson, 1981; Grossberg, 1982, 1987a; Kohonen, 1989a), and neurocomputing (Anderson and Rosenfeld, 1988) has given a new impulse to the study of learning in information processing models. The quasi-neural elements and syntax of the connectionist language provide a useful formalism that is very well suited for incorporating learning abilities without, however, restricting the field to the extent that it could be called a unitary theory. Just like their biological example, some learning networks may self-organize by selecting and categorizing relevant information, and by retaining this over time. It is hoped that by merely placing a learning network in an environment implicit 'programs' will emerge as a consequence of the interaction of the network with this environment.

Although the importance of learning has been recognized in the behavioural sciences for a long time (e.g. Ebbinghaus, 1885; James, 1890), it has been relatively neglected in conventional cognitive models and classical artificial intelligence systems, based on the computer metaphor. These models are often based on the assumption that it is possible to define and to formalize all necessary knowledge, and to incorporate this knowledge in a model. Such a view, of course, limits the knowledge that can be represented in an information processing model to the knowledge that can be made explicit by the designer.

3

Moreover, the presupposition that all knowledge can be implemented in the form of well-defined rules is neither proven nor probable (see Wolters and Phaf, 1990). Pattern recognition, for example, which requires the simultaneous processing and integration of a large amount of information, is an ability that the human system can perform with ease, but has proven an exceedingly hard task for most conventional, rule-based approaches. Even at this stage of development, learning neural networks perform functions for which no computing algorithm has been found. The robot arm developed by Kuperstein (1988), for instance, learns to direct its arm to objects detected by its two video cameras, a task that so far has defied a rigorous analytic solution.

The idea of non-formalizable knowledge that can be learned by, but not programmed into, an information processing system may have profound consequences for the practice of connectionism. Approaches to neural networks, that want to describe the knowledge represented in a network in a mathematical fashion, may restrict their models to functions that bear little relevance to the functions that characterize the natural systems. If easily formalizable functions were to exist, a hundred years of experimental behavioural research would probably have yielded such general functions. In fact, it can be argued that, when a model can be completely formulated in the mathematical language, a connectionist formulation is not necessary. Although mathematical analysis may be an important goal for connectionist research, we adhere to the standpoint that psychological plausibility of neural network models should take precedence over rigorous mathematical tractability. In this work a new learning network model will be presented starting from some more psychologically and biologically oriented considerations.

1.2 Some problems with learning neural networks

Approaches to learning information processing systems are still in their infancy, and despite preliminary successes, most of the currently popular learning network models show a number of shortcomings and problems. Among the problems are lack of speed, lack of stability, inability to learn either with or without supervision, and inability both to discriminate between and generalize over patterns.

A lack of speed, in particular, is hampering completely structureless, homogeneous models and models that assume a hierarchical, layered structure with total connectivity between nodes of adjacent levels. If the size of the models increases, the quadratic rise in the number of modifiable connections may lead to a prohibitive lengthening of the time to reach stable states (e.g. Perugini and Engeler, 1989). The network by Rumelhart *et al.* (1986), for instance, already needs hundreds to thousands of presentations to learn even a simple function such as the EXOR. Though many improvements of the backpropagation procedure have been proposed, lack of speed still remains a major problem for this class of networks.

Total homogeneous connectivity may also result in a lack of stability of representations, because there is too much susceptibility to interference. Every input–output relation that can be learned by such a network has an *a priori* equal status to all other possible relations. Every new relation can and will, therefore, interfere with every old relation, if the latter is not strengthened over and over. The problem of 'catastrophic interference' in neural networks has been discussed by McCloskey and Cohen (1989) and by Ratcliff (1990). Much recent work is aimed at investigating this problem (see Chapter 7 for an overview). They performed extensive tests on the backpropagation procedure and showed that well-learned information is replaced rapidly when new information is learned and the old information is not presented again. If such a network has learned a stimulus set *A* to perfection and is subsequently trained on another set *B*, it may be able to learn the set *B* to perfection as well, but in the course of training it will forget most of set *A*. One way to accomplish perfect learning of both sets is to retrain over and over with both *A* and *B*. Several extensions of backpropagation have recently been proposed to deal with this issue. It appears that full connectivity is only one of the problems causing massive interference (see Chapter 7 for a more detailed discussion of this problem).

Another problem with the delta-rule and backpropagation learning procedures is that they only allow for supervised, but not unsupervised, learning. When these networks are said to function without supervision, usually an autosupervision scheme is used, where input and desired output pattern are the same. Of course, from a psychological point of view, supervised learning (i.e. with instruction and correction of errors) may also be important. Indeed, even in many cases without apparent explicit supervision, such as uninstructed skill learning and operant conditioning, there may still be autosupervision. The result of some action may be compared with some internal standard and the action may be corrected accordingly. Yet, much learning, like the incidental storage of everyday experiences, proceeds without any form of supervision and the inability of these learning procedures to handle such learning seems to be a severe shortcoming.

As far as unsupervised learning is accomplished by using some form of Hebbian-type learning rule most models are capable of learning by autonomously discriminating between different input patterns. Some others allow for generalization over similar input patterns as may be necessary for the recognition of constancies, like invariance for size, and translation as, for instance, in handwriting. However, none of the existing models seems to combine both abilities in an efficient manner. For example, the well-known adaptive resonance theory (ART) (Carpenter and Grossberg, 1987; Grossberg, 1976) is capable of further and further discriminations, but it may have only a limited capacity to cluster patterns under a common representation. Patterns that do not sufficiently match any of the learned representations may be rejected by an ART module that is completely 'filled' with other representations.

So, it seems that most currently available learning networks have architectural characteristics, or use learning rules, that result in a variety of problems and shortcomings in comparison with the human system. In part, this may be caused by the fact that neural network research seems to have been guided primarily by the availability of computational and mathematical methods, rather than by biological or psychological constraints. It will be argued here that, if the current connectionist language is supplemented with new terms derived from psychology and the neurosciences, the combined effort may lead to more plausible network models and may help to solve some of the above problems.

1.3 Structural constraints

1.3.1 Limitations on connectivity: modularity

Many of the currently popular network architectures show little, if any, structural constraints. Some networks assume total interconnectivity between all nodes (e.g. Hopfield, 1982). Others assume a hierarchical, multi-layered structure (e.g. Rumelhart *et al.*, 1986) in which each node in a layer is connected to all nodes in neighbouring layers.

Completely connected architectures allow virtually any possible input–output relation (Funahashi, 1989; Hornik *et al.*, 1989; Stinchcombe and White, 1989) to be learned. They exhibit extreme plasticity. The position taken by these connectionists in the nature–nurture debate seems to be at the far 'nurture' end. Psychological evidence, however, indicates that for the human system such an extreme position may not be warranted. Some learning tasks seem to be much easier than others (e.g. learning a motor skill with the preferred or the non-preferred hand). Complete connectivity also suggests that any multi-task execution would be hampered by mutual interference. But results from interference studies in humans show that many tasks can be performed simultaneously without mutual interference (e.g. speaking while driving a car, or perceiving and producing speech; Shallice *et al.*, 1985), whereas other task combinations are almost impossible to perform at the same time, especially if some task elements are shared (e.g. listening to two conversations at the same time (see also Allport, 1980)). Furthermore, there is a large body of neuropsychological evidence showing that isolated abilities, such as the ability to recognize faces (e.g. Damasio *et al.*, 1982), or to speak fluently, may be lost without affecting other cognitive abilities in any way (e.g. Gazzaniga, 1989; Luria, 1973; Shallice, 1988). In summary, these data seem to indicate that the human information processing system consists of modules – relatively isolated subsystems – that can function quite independently of each other.

Also, neuroanatomy provides a wealth of evidence showing that the human brain does not have total and uniform connectivity. For one thing, this would require 10^{11} connections per neuron, whereas only about 10^4 connections are available. On a macroscopic scale many structurally different cortical and subcortical centres can be distinguished that are only partly interconnected (Kosslyn *et al.*, 1990; Livingstone and Hubel, 1988; Zeki and Shipp, 1988). On a microscopic level, the minicolumns found in the grey matter of the neocortex can be considered module-like structures (e.g. Mountcastle, 1978; Szentágothai, 1975). The minicolumns consist of small regions in the neocortex having intracolumnar connections that are for a large part inhibitory, and long-range afferent and efferent excitatory connections to other neocortical columns via the white matter and subcortical centres (e.g. Creutzfeldt, 1977).

In a connectionist model, brain-style modularity may provide a coarse initial architecture on which learning imparts a finer structure (e.g. Changeux and Danchin, 1976). The initial architecture determines what can and what cannot be learned. Though it is improbable that the genes code all structural information about the brain (Changeux and Danchin, 1976), it also seems highly unlikely that this global modular structure is completely determined by the learning history of the organism. The neuroanatomical structure itself is highly regular, not only within but also between individuals. Many brain-damage effects are about the same for all individuals and these are not altered very much by experience. Also, characteristic early development of animals and humans must − at least partially − be the result of neuroanatomical constraints.

Architectural constraints are in many cases necessary, rather than merely advantageous. This can be illustrated when we consider a learning version of the McClelland and Rumelhart (1981) interactive activation model for context effects on letter recognition, which is basically a modular model. In the vertical direction, the four positions of the letters and features are separated up to the word level. If the representations of this model had to be learned with, for instance, a layered backpropagation network, there would have been no way in which the position information would have remained separate up to the third level (see also Figure 1.1). The modules for the four positions at the letter level have to be there from the start, because without the modular constraints a letter module at, say, the first position could in the course of learning easily become connected to features at the second, third and fourth positions. This would yield a model with totally different characteristics. In a similar vein, we could argue that the nervous system must constrain the formation of cortical connections to prevent undesirable interactions among brain structures, for example the visual system giving rise to illusory auditory sensations after prolonged training with paired visual and auditory stimuli.

A modular architecture provides other advantages as well. First, a modular architecture may lead to an increased stability of representations and a reduced interference by subsequent learning, because of the reduced plasticity. It has been shown that the introduction of modular constraints in a layered architecture may result in faster learning and better solutions (e.g. Rueckl *et al.*, 1989). Secondly, modular architectures generally lead to a reduced number of connections per node. The reduction in the number of connections also provides an important practical advantage because it makes hardware implementations of neural networks easier. A limiting factor in many hardware

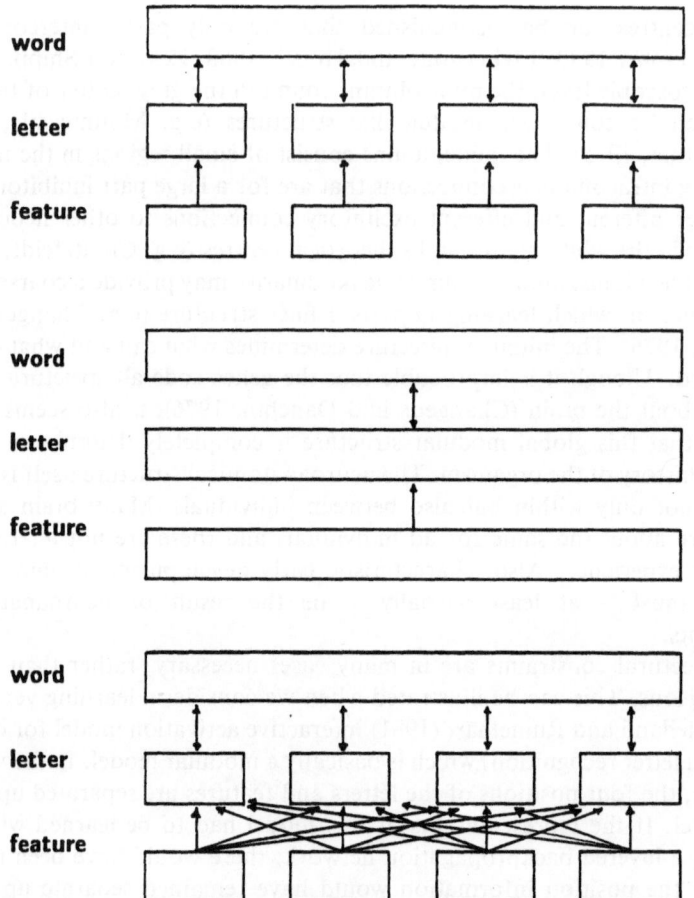

Figure 1.1 Top: a schematic drawing of the interactive activation model for letter recognition (McClelland and Rumelhart, 1981). Middle: a non-modular version of the model prior to learning in a supervised mode. Bottom: the probable result of learning the model with only layers as initial structure. Spurious connections between features and letters may be formed.

implementations has been the unmanageable increase in the number of connections with the number of nodes in a homogeneous network. Finally, in many practical applications of difficult tasks it is desirable to introduce *a priori* knowledge to guide learning in the intended direction. Prestructuring networks in a modular way is one possible way of achieving this.

1.3.2 The organization of excitatory and inhibitory connections

A structural characteristic often found in neural networks which seems to be rather implausible is the apparent random organization of inhibitory and excitatory connections. It may result from changes in sign of connection weights due to learning. Negative weights (i.e. inhibitory connections) may become positive (i.e. excitatory connections) and vice versa. Neurophysiologically this is rather unlikely. It is in conflict with Dale's law (e.g. Eccles, 1957; Kandel and Schwartz, 1985) which implies that a neuron can give off only one kind of synapse: either inhibitory or excitatory (see also Crick and Asanuma, 1986). Therefore, further architectural constraints should be carried through prohibiting any change in sign of a connection weight and restricting all output connections of a given node to either the excitatory or the inhibitory type.

An effective way to implement this constraint, and to combine it with modularity, is to use the general architecture of the neocortical minicolumns as the basic design principle of a connectionist module (Szentágothai, 1975). These modules are about 0.2–0.3 mm in width and 2.5–3 mm deep (i.e. they extend through the entire human cortex). The modules consist of excitatory pyramidal cells (about 60%) and various types of inhibitory interneurons, such as basket cells. The pyramidal cells in the upper layers of the cortex mainly connect to other cortical regions in the same hemisphere; those in the deep layers generally connect to subcortical centres (Kandel and Schwartz, 1985). The pyramidal cells may thus form long-range excitatory connections. The interneurons, which are prevalent in the middle layers, mainly have short-range inhibitory connections. For the cerebellar cortex a comparable functional organization for the connections can be found (Blomfield and Marr, 1970; Keeler, 1988). Though the division into modules seems to be less clear for this part of the brain, the functional distinction between the long-range and short-range connections may be similar to that of the neocortical functional module. Given such an architecture, the principle that governs intermodular organization seems to be excitation, and the principle governing intramodular organization seems to be inhibition.

The structural principle of intramodular inhibition implies that the main process within a module will be competition. Competition is a powerful mechanism that has been used to simulate a large number of psychological and neural phenomena ranging from masking phenomena to learning processes (e.g. Bridgeman, 1971; Cornsweet, 1970; Fukushima, 1975, 1988; Grossberg, 1982, 1987a; McClelland and Rumelhart, 1981; Rumelhart and Zipser, 1985; Phaf *et al.*, 1990b; Von der Malsburg, 1973; Walley and Weiden, 1973). It enables a system autonomously to categorize new patterns, and it can be used, therefore, to implement unsupervised, competitive learning.

1.4 Functional constraints: attention and learning

Connectionist models exhibit an extreme form of parallel processing. No central processor, or executive system, is available to control the information flow through a network. Instead, control of processing in a network is distributed over all nodes and can be seen as an emergent property of the collective behaviour of the nodes. In the absence of a central controller it is hard to see how networks can distinguish between the significant and the trivial, between novel and familiar inputs, which seems to be a key property of the human information processing system. In psychological terms they lack the ability to attend to relevant stimuli and neglect irrelevant stimulation. Recently, Phaf *et al*. (1990b) have shown how task-specific selective attention can be modelled in a modular network.

In addition to a selection function, attention also seems to have an arousal function. Hebb (1955), for example, postulated that a stimulus has both an (informational) cue effect and a non-specific (non-informational) arousal effect. There is some neurophysiological evidence for such an arousal process triggered when new or unexpected stimuli are encountered. For instance, it appears that non-specific activating signals are sent to higher cortical levels from lower subcortical centres, like the reticular formation and hippocampus–amygdala complex (e.g. Halgren *et al*., 1980), possibly after feedback from the higher levels. It is interesting to note that the hippocampus and amygdala presumably play an important role in the learning mechanism because damage to these structures is strongly associated with anterograde amnesia (e.g. Mishkin, 1978). Also, other authors have made a similar connection between attention, learning and specific arousal processes in the brain (e.g. Luria, 1973; Murray and Mishkin, 1984). Several researchers, notably Grossberg (e.g. 1982 and 1987a), have expressed the need for such an *attentional–arousal mechanism* in neural networks and have developed such mechanisms. The development of the adaptive resonance theory is partly an answer to what Carpenter and Grossberg (1988; Grossberg, 1982, 1987b) have called the stability–plasticity dilemma: 'How can a learning system be designed to remain plastic, or adaptive, in response to significant events and yet remain stable in response to irrelevant events?' (e.g. Carpenter and Grossberg, 1988, p. 77). Plasticity is necessary for the incorporation of new representations in a network. Stability means keeping old representations intact.

A potential solution to the stability–plasticity dilemma could be obtained if an attentional mechanism could be implemented that distinguishes between old and new representations, and that uses this information to control the learning process. New representations must be encoded quickly, which calls for high plasticity or, in other words, for a high learning rate. Old representations necessitate only slow learning to improve their stability further. A high learning rate with old representations would only damage other representations, causing unnecessary interference. In memory psychology such a distinction between two forms of learning has been made by Graf and Mandler (1984; see also Mandler, 1980). They distinguish between *elaboration* learning and *activation* learning. Elaboration learning results in the formation of new associations, activation learning merely strengthens existing associations. Graf and Mandler have applied this distinction to explain dissociation phenomena in explicit and implicit memory tests (for a review, see Schacter, 1987).

According to Graf and Mandler, tests for *explicit* memory, like free recall and recognition, rely on the conscious recollection of the event that is being remembered. These tests require elaboration learning, because new associations between stimuli and the context in which they occurred have to be formed (e.g. Raaijmakers and Shiffrin, 1981). Tests for *implicit* memory, like word completion and threshold identification, show a facilitation of task performance by previous presentation but do not require the explicit recollection of presentation context. For these tests, activation learning is sufficient because all that is needed is the strengthening of associations, which allows a faster or more efficient future access to the already stored representation of the stimulus.

In order to incorporate these learning processes in a neural network, it would have to shift towards elaboration learning with novel input activation patterns, and towards activation learning with old inputs. Elaboration learning would call for an exploratory process that finds possible representations in the network, and an increased learning rate to encode quickly the newly found representations. Activation learning would only involve activation of existing representations and base rate learning to strengthen them further.

1.5 Implementation of the constraints

In order to implement the attentional and learning processes discussed in the previous section the intramodular competition principle discussed earlier can be used. Intramodular competition results in local representations (representations on single nodes, or semi-distributed representations, see Chapter 7). Initially, presentation of an activation pattern has the effect of activating many nodes. When a new pattern is presented, there will be no *a priori* fitting node available and several nodes will become activated simultaneously. Because we assume that these nodes have strong mutually inhibitory connections, competition within the module may be fierce. Learning has the effect of reducing competition in a module. Repeated presentation of a firmly learned activation pattern will strongly activate its representational node. Because it will no longer activate other nodes, however, there will be little competition. Recall that competition serves two purposes: first, it should lead to an increase of arousal (to support the search for a new representation) and, secondly, it should increase learning speed (to incorporate the new representation in the weight pattern quickly).

In our model the arousal effect of competition is implemented by the application of random activations to nodes in the module with an amplitude that depends upon the amount of competition, which, in turn, depends upon the novelty of the activation pattern to a module. Following Phaf *et al.* (1990b) attention is viewed as the selective enhancement of activations (see also Spitzer *et al.*, 1988). The selection should of course not be made beforehand, but has to be performed by the model. The attentional activations must, therefore, be aspecific. The random activations lead to higher but differentiated activation values of the competing nodes which helps in solving the competition and the search for a new representation. The faster learning rate, which is necessary to incorporate quickly a new representation into the network, is implemented by introducing a variable in the learning rule with a value that is a function of the amount of competition. This will cause learning speed to be low when already represented inputs are presented again (little competition), whereas learning speed increases when not yet represented inputs are given (much competition). The reduction in learning rate and random activations with repeated presentation of a pattern may be compared with habituation responses found in all organisms. Presentation of an unfamiliar stimulus to an animal gives rise to an arousal response, which may be accompanied by an orientation reaction (Näätänen, 1986; Sokolov, 1960, 1966, 1975). But repetition of the same stimulus results in a gradual habituation of the response.

The introduction of noisy activations supporting the exploratory process, necessary for finding a new representation for novel patterns, performs a similar function as the 'quasi-thermal' (temperature-dependent) fluctuations used in the Boltzmann machine (e.g. Ackley *et al.*, 1985). In the Boltzmann machine such fluctuations are used for escaping from suboptimal solutions, for example from the activation of representations that do not correspond maximally (in terms of some cost function) to stored representations. The aspecific activations in the Boltzmann machine cannot be identified with the arousal function postulated by Hebb (1955), because the activations are not dependent upon the stimulation, but follow a fixed 'annealing' or 'cooling' scheme. Recently, however, Lewenstein and Nowak (1989a,b) have proposed a 'self-induced simulated annealing method' in which the noise level is also made dependent upon the novelty of input patterns. These state-dependent fluctuations, however, are limited exclusively to the enhancement of the retrieval of patterns and do not play a role in the storage of patterns. In the procedure outlined in this work, however, the aspecific random activations are state dependent both in the storage and in the retrieval phase. In fact, there is no fundamental distinction between storage and retrieval. By definition, old patterns are presented during retrieval. Hence, retrieval involves activation learning whereas elaboration learning predominates during storage.

Random activations have three important effects. Firstly, they prevent competition deadlocks when competing nodes are activated to the same degree. Secondly, random activations make it possible to escape from shallow attractors (e.g. Hopfield and Tank, 1986) and to reach deeper ones. They ensure more optimal solutions in the search for a suitable representation. Thirdly, once a module has arrived at a categorization (i.e. has assigned a representation node to an activation input pattern) the random activations, which will be strongly suppressed in all nodes but the winning node, will increase the activation of the winning node. This has the effect of enhancing the contrast between the activations of the 'winning' and 'losing' nodes.

The introduction of competition-dependent learning rate models the difference between elaboration and activation learning. As mentioned earlier, the differences between elaboration and activation learning have been used to explain dissociation effects between explicit and implicit memory. Moreover, there is much support in the psychological literature for a positive relation between arousal and (long-term) retention (e.g. Kleinsmith and Kaplan, 1963; Phaf and Wolters, 1986; for reviews see Eysenck, 1982, and Hockey et al., 1986). Unlike all other mechanisms discussed so far, the control of learning as proposed here is not strictly local. It operates at the level of the module rather than at the level of a single node. The mechanism may be compared with the spreading of some neuromodulator, its concentration determining the amount of possible change in synaptic efficiency. Recently NMDA has been suggested as a likely candidate for such a process (e.g. Kleinsmidt et al., 1987; Brown et al., 1988), although other neuromodulators may perform this function as well (e.g. Bear and Singer, 1986).

An important choice to be made in any learning network model concerns the learning rule. We have adopted a modified version of the rule introduced by Grossberg (1976; see also Carpenter and Grossberg, 1987). In order to vary learning speed as a function of the amount of arousal (or competition) in a module, an extension of the rule was introduced by making the learning parameter (comparable with the constant of proportionality in Hebb's rule (Hebb, 1949)) dependent upon the amount of competition.

An important difference between the Grossberg rule and the Hebb rule concerns the fact that in the former the weight change of any connection is a function of the activations of all nodes contributing to the activation of the receiving node, whereas in the latter only the node sending activation through the specific connection is taken into account. As has been shown by Carpenter and Grossberg (1987) this makes the Grossberg rule capable of distinguishing between correlated patterns through what they call a 'Weber law' rule. We made the learning rule somewhat more local than the original Grossberg rule by making the weight change dependent upon the sending activations multiplied by the weights to the receiving node. In this manner, only the effective (weighted) activations, which are locally available in a receiving node, are used in the modification of a weight. Further differences between the Hebb rule and the rule proposed by Grossberg concern the asymptotic approach of minimum and maximum weights and the asymmetry between sending and receiving activations. This asymmetry results in much larger changes in weights to 'winning' nodes, than to 'losing' nodes.

A final point to be mentioned in the use of the learning rule is that it applies only to excitatory connections. More specifically, only excitatory intermodular connections are assumed to be modifiable through learning, whereas all intramodular connections (both inhibitory and excitatory) are assumed to remain fixed. The reason for this constraint is that excitatory intermodular connections are especially important in combining information from different sources, and because only these connections carry information about the environment. These connections form the informational constraints that have to be satisfied in the retrieval of stored information. In contrast, intramodular inhibitory connections only play a role in the execution of the selection mechanism governing the winner-take-all competition in a module. These connections do not carry specific content information; they are only instrumental in the constraint satisfaction process. There is no need for these connections to adapt to external information.

Chapter 2

Description of CALM

2.1 Structure of CALM

2.1.1 Elements and activation

We can now join together the different aspects of our view on learning networks in order to construct the basic module: CALM, the Categorizing And Learning Module (see also Murre *et al.* (1989a,b, 1990a,b,c, 1992) and Vogl (1989)). This module may serve as a basic building block for larger networks performing hierarchical categorization and clustering. The architecture of such a network is determined by interconnecting specific modules and by not interconnecting others. In this way, different global network architectures can be designed for the specific functions a network will have to perform.

In the following sections we shall present the basic design features of a single CALM module, and we shall discuss and illustrate some of its functional characteristics. CALM has a fixed structural design consisting of a certain arrangement of the nodes and their interconnections. A *node* is a simple

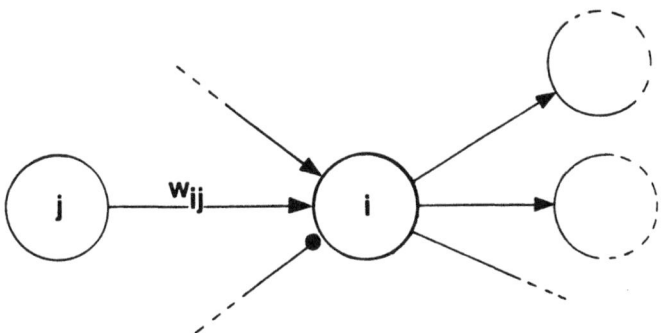

Figure 2.1 Nodes i and j, with a connection w_{ij} from node j to node i. Throughout this book arrow heads indicate excitatory connections (either modifiable or fixed), full circles indicate inhibitory connections. Reprinted with permission from *Neural Networks*, vol. 5, Murre, J.M.J., Phaf, R.H. and Wolters, G., CALM: Categorizing and learning module, Copyright 1992, Pergamon Press Ltd.

processing element that manipulates a unidimensional variable, called *activation*, according to a small number of fixed rules. The activation of a node is a real-valued number between zero and one. Every node receives and integrates input from a number of nodes connected to its input side. The resulting activation value is transmitted over connections fanning out from the output side (see Figure 2.1).

The effectiveness of the activation transmission through the connections is expressed in a real-valued number called the (connection) *weight*. The effective input to node i, called excitation e_i, is the weighted sum of the individual activations of all nodes connected to the input side of the node. The input may be either positive (excitatory) or negative (inhibitory). The activation of node i at time $t + 1$, denoted as $a_i(t + 1)$, is a function of its activation at t, $a_i(t)$, and its input excitation e_i. This function is expressed by the following rule:

excitatory input:

$$a_i(t + 1) = (1 - k)a_i(t) + \frac{e_i}{1 + e_i} [1 - (1 - k)a_i(t)], \quad e_i \geqslant 0 \qquad (2.1a)$$

inhibitory input:

$$a_i(t + 1) = (1 - k)a_i(t) + \frac{e_i}{1 - e_i} (1 - k)a_i(t), \qquad e_i < 0 \qquad (2.1b)$$

where

$$e_i = \sum_j w_{ij} a_j(t)$$

w_{ij} denotes the weight of a connection from node j to node i (see Figure 2.1), $(t + 1) - t$ is a unit time interval, called the iteration time, and k, the decay factor, is a constant between zero and one. The time units will not be specified here and may, in fact, be taken infinitesimally small, in which case the above difference equation will become a first-order differential equation.

In the activation rule three components may be distinguished. The first component, $(1 - k)a_i(t)$, represents the autonomous decay of the activation of a node. If no input is present ($e_i = 0$) the node's activation simply decays to zero with a rate that is determined by k. In a continuous time representation this decay will be exponential. Therefore, k must be chosen between zero and one. The second part, $e_i/(1 + e_i)$, 'squashes' the input excitation to a number between zero and one. The third part of the rule, $[1 - (1 - k)a_i(t)]$, for $e_i \leqslant 0$, ensures that the increase in activation due to input excitation diminishes as the activation approaches the maximum activation. This causes an asymptotic approach to the maximum activation value. Similarly, for $e_i < 0$, $e_i/(1 - e_i)$ squashes the negative excitation (inhibition) between minus one and zero. The $(1 - k)a_i(t)$ component ensures the asymptotic approach to the minimum activation value.

2.1.2 Internal structure

Functional categories of nodes

Though all nodes in the CALM module have identical activation rules, it is possible to classify them into three different, mutually exclusive, categories. The classification is based on the nature of the connections to other nodes, which in turn determines their function. The internal connections in a CALM module are fixed and will be described in the next section.

The first category consists of nodes that have modifiable connections to similar nodes in other modules and fixed connections to nodes in the same module. All outgoing connections of these nodes are excitatory and have positive values. Because the activation of such a node may correspond to the presence of a particular pattern of input activations to a module, it is called R-node (from Representation-node). R-nodes are the only nodes in a CALM module that may send or receive excitation to or from nodes in other modules.

Nodes in the second category are called V-nodes (from 'Veto-nodes', see Crick and Asanuma, 1986). V-nodes only have inhibitory (negative) outgoing connections. A V-node inhibits all other nodes in a module. In particular, V-nodes strongly inhibit each other, so that competition arises among the V-nodes. Every V-node receives excitatory input from only one R-node. The R-nodes thus form matched pairs with the V-nodes (see Figure 2.2).

The third functional category consists of only one node, which is excited by all R-nodes in a module, and inhibited by all V-nodes. It is called the A-node (from 'Arousal-node'). Because of the specific wiring pattern of the connections (see Figure 2.2) the activation of the A-node is a positive function of the amount of competition in a module. In a CALM module competition will be most prevalent for new input patterns that have not been learned yet, and that activate several uncommitted R-nodes. The A-node activation is, thus, a measure of the novelty of the input pattern.

In Figure 2.2 the module is shown with the lower row consisting of R-nodes and the upper row containing V-nodes. The A-node is positioned on the right-hand side of the module. The functional characteristics of the module are not dependent upon the size of the module (i.e. the number of V- and R-nodes). The number of V- and R-nodes, therefore, is not specified in the general design in the figure.

In Figure 2.2 still another kind of node appears, which has been located outside the module. Because of its location it is called E-node, from 'External-node'. The node is thought to correspond to some part of an arousal centre outside the module. In the nervous system there are a number of subcortical centres that diffusely activate large parts of the cortex. Several E-nodes may be combined to form such a centre. In most of our models we do not actually interconnect these nodes. So, generally, an E-node has connections to only a single module. The E-node receives input from the A-node in this module, and in turn sends random activation pulses to all R-nodes in the module. These random pulses are uniformly distributed over the interval $[0, a_E(t)]$, where

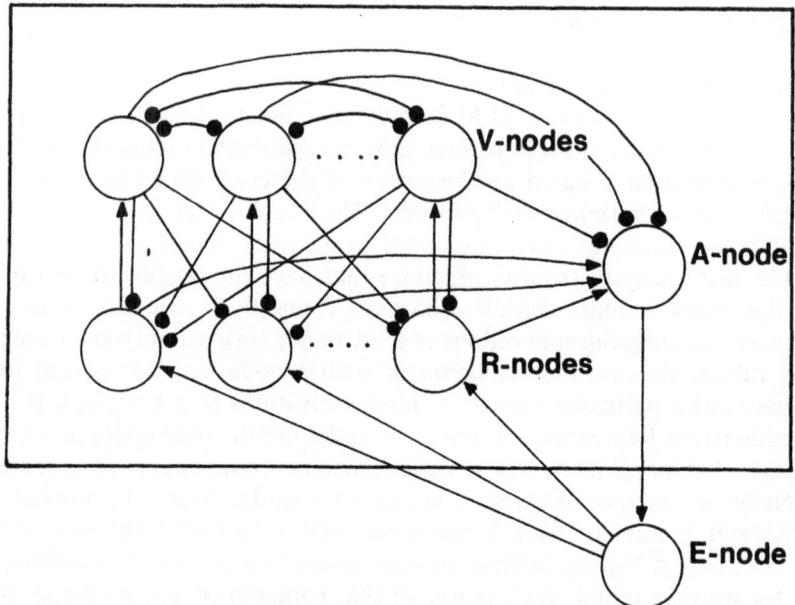

Figure 2.2 Schematic drawing of the internal wiring pattern of a CALM module. Shown are the three internal node categories V-nodes (Veto-nodes), R-nodes (Representation-nodes) and A-node (Arousal-node), as well as the E-node (External-node). Reprinted with permission from *Neural Networks*, vol. 5, Murre, J.M.J., Phaf, R.H. and Wolters, G., CALM: Categorizing and learning module, Copyright 1992, Pergamon Press Ltd.

$a_E(t)$ stands for the activation of the E-node at time t. The E-node also controls the learning rate in a module. This aspect will be dealt with in Section 2.1.3.

Categories of internal connections
It is convenient to classify all connections in a module into a number of mutually exclusive categories and give them names that can be referred to in the text. The weights of all intramodular connections are non-modifiable. The values of the internal weights are real numbers that are not necessarily restricted to a particular interval. (In this work values in the range $[-10,3]$ have been used.) The values of the modifiable, intermodular weights that connect different modules are restricted to the interval $[0,K]$, where K is a positive constant, for instance 1.0 or 2.0. Modifiable or learning weights are discussed in the next section.

Figure 2.3 offers a more detailed view of the module with the various connection types indicated. A brief description of each of the eight connection

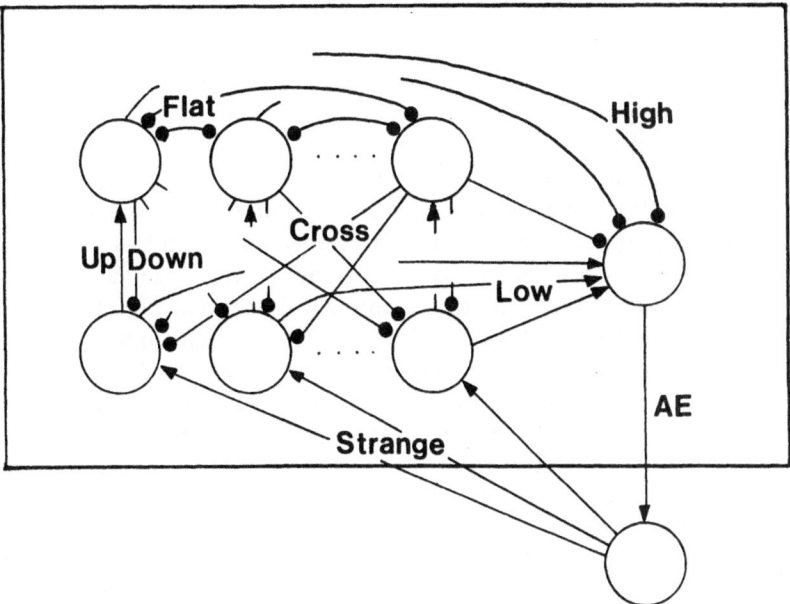

Figure 2.3 Schematic drawing of the wiring pattern as in Figure 2.2, but now with connection categories indicated. Reprinted with permission from *Neural Networks*, vol. 5, Murre, J.M.J., Phaf, R.H. and Wolters, G., CALM: Categorizing and learning module, Copyright 1992, Pergamon Press Ltd.

types follows:

1. *Up weight* (excitatory): connects an R-node to a V-node. A particular V-node is paired to a particular R-node by this connection.
2. *Down weight* (inhibitory): is the reciprocal connection to the up weight. The up and the down weight together cause an R–V-node pair to exhibit differentiating characteristics with respect to changes in activation input to the R-node.
3. *Cross weight* (inhibitory): controls recurrent lateral inhibition from V-nodes to R-nodes. These weights are usually strongly negative, so that a V-node may veto all non-matched R-nodes.
4. *Flat weight* (inhibitory): controls competition between V-nodes and, thus, between the R–V-node pairs.
5. *Low weight* (excitatory): controls excitation of the A-node by the R-nodes. Many activated R-nodes tend to increase the activation of the A-node.
6. *High weight* (inhibitory): controls inhibition of the A-node by the V-nodes. Many activated V-nodes tend to decrease the activation of the A-node.
7. *AE weight* (excitatory): regulates the influence of the A-node activation on the activation of the E-node.

8. *Strange weight* (excitatory): regulates the amplitude of random activations from the E-node to the R-nodes.

All weights of a particular type have equal values. The values of the weights together with the parameters of the activation and learning rule (see next section) completely determine the properties of the module. (See Appendix A for a complete list of the values used throughout this work.) For some weights it is possible to deduce (relative) values given a certain desirable property of the module (see also Appendix A). For instance, since we want a V-node to inhibit strongly all R-nodes (except the paired R-node) a very large value has been chosen for this weight type. Another example is the ratio of high/low weights. The low weights excite the A-node, whereas the high weights inhibit this node. For the proper functioning of the module (to be discussed below) it is important that, as soon as the competition has been solved and a single R–V-node pair has become activated, the activation of the A-node decreases to zero. From this it can be deduced that the absolute value of the inhibitory high weight must be larger than the value of the excitatory low weight. In some cases the relation between module properties and weight values is more complicated. For these cases the weight values were found by a process of heuristic parameter estimation.

2.1.3 External structure: learning connections

In general, full connectivity is assumed between connected modules at different levels. This means that, if there is a connection to module 2 from module 1, every R-node in module 2 receives input from all R-nodes in module 1. The weights of these connections are called *interweights*, because they are concerned with intermodular connections. All interweights are modifiable, with possible values in an interval $[0,K]$, where K, a positive constant, is a parameter of the learning rule (see below). They are adjusted according to a variation of Grossberg's (1976) learning rule, which is related to the Hebb rule (Hebb, 1949). In the Hebb rule a weight change is dependent upon the correlation between the activation of sending and receiving nodes. Grossberg (1976) modified this rule by also taking into account the total background excitation caused by 'neighbouring' R-nodes of the R-node sending activation through the connection that is being changed (see Figure 2.4). Another modification concerns the fact that the weights can only approach their minimum value (0) and maximum value (K) asymptotically. We added two further changes which were discussed earlier (see Section 1.4). The first is that we have made the learning rate μ_t dependent upon the novelty of the pattern. The second change is that we take into account the *weighted* background activation, $\Sigma w_{if}(t)a_f$, whereas Grossberg (1976) uses the *unweighted* background activation, Σa_f. If the unweighted activations are used, a conceptual problem is how these become known to the receiving node. Clearly, some non-local influence must

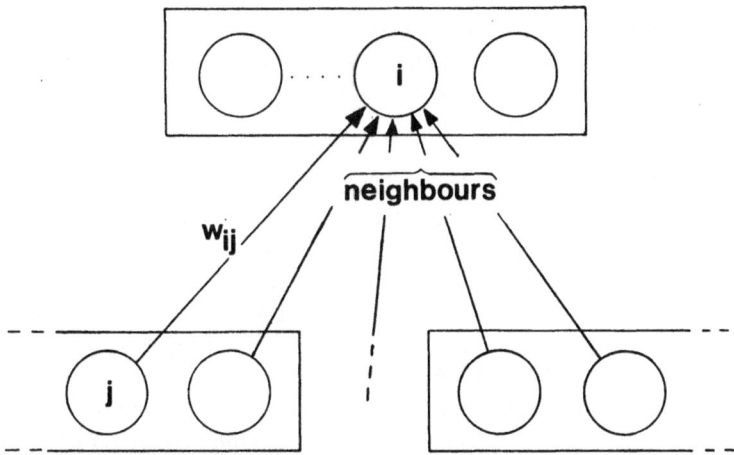

Figure 2.4 Illustration of neighbour activations. Neighbouring R-nodes and connections are relative to the connection under modification. w_{ij} indicates the weight that is being changed by the learning rule. The weighted neighbour activations present background activations. All nodes shown are R-nodes. For simplicity, other types of nodes have been left out. Reprinted with permission from *Neural Networks*, vol. 5, Murre, J.M.J., Phaf, R.H. and Wolters, G., CALM: Categorizing and learning module, Copyright 1992, Pergamon Press Ltd.

be postulated to solve this. By using the weighted activations instead, which are readily available at the node, the rule better preserves the principle of locality.

In the resulting learning rule, a_f, a_i and a_j stand for $a_f(t)$, $a_i(t)$ and $a_j(t)$, respectively; $w_{ij}(t)$ is the interweight between R-nodes j and i (from j to i), $w_{if}(t)$ indicates an interweight from a 'neighbouring' R-node f (of j) to R-node i, $\Delta w_{ij}(t+1)$ is the change in weight from j to i at time $t+1$:

$$\Delta w_{ij}(t+1) = \mu_t a_i \left([K - w_{ij}(t)] \, a_j - L w_{ij}(t) \sum_{f \neq j} w_{if}(t) a_f \right) \qquad (2.2)$$

Note that f, i, and j must be R-nodes. L and K are positive constants. K determines the maximum value of an interweight, which may be approached asymptotically by w_{ij}. The first term within the large parentheses is always positive and represents increases in the weight. The second term is responsible for all decreases in the weight. An inactive connection ($a_j = 0$) to an activated node (a_i) will always decrease, because only the second term will be non-zero. The increase/decrease ratio can be controlled by the K/L ratio. Furthermore, μ_t represents the Hebb parameter, controlling the learning rate in the module, and is equal to

$$\mu_t = d + w_{\mu E} a_E \qquad (2.3)$$

where d is a constant with a small value (from the E-node to the learning parameter), and a_E is the activation of the E-node. Because an interweight should be confined to the interval $[0,K]$, a complete description of the learning rule must include the following condition:

$$w_{ij}(t+1) = \max\{\min[w_{ij}(t) + \Delta w_{ij}(t+1), K], 0\} \qquad (2.4)$$

That is, the weight values are kept within the allowed range. This boundary condition, however, cannot be violated with sufficiently small values of μ_t.

In the learning rule departures from the Hebb rule are that high-weight values at time t tend to limit the increase of the weight, and that μ_t is made dependent on the activation of the E-node. The second term within the large parentheses in (2.2) can be seen to form an extension to the Hebb rule, which was first used by Grossberg (1976). In the case of a high background excitation, $\Sigma_{f \neq j} w_{if}(t) a_f$, increases in weight will become smaller and weights may even decrease, especially if $w_{ij}(t)$ is high as well. Thus, the effect of this component is an adaptive downscaling of all weights, whenever the total input to the node will be too high. This may happen, for instance, if many modules are connected to one module, or if high input activations are present. In this manner, the Grossberg learning rule automatically normalizes the modifiable weights to a level suited for the sizes of the modules they connect. The downscaling provides a mechanism that prohibits an overall increase in weights to the same saturation level which eventually might paralyse the entire system. There are still other reasons for including this downscaling part in the learning rule. This has been discussed by Carpenter and Grossberg (1987) within the framework of their 'Weber law rule'. It plays a very important role in the discrimination between correlated, non-orthogonal patterns, which will be described in more detail later on (see Sections 2.2.3 and 2.2.4).

2.2 Functioning of CALM

2.2.1 Excitatory, inhibitory and arousal processes

In order to understand the dynamics of a CALM module it is useful to distinguish three processes that determine its working: the excitatory, the inhibitory and the arousal process (see Figure 2.5). The excitatory process, consisting of the activation of the R-nodes in a module, is triggered by stimulation of inputs to the module. The stimulation feeding the excitatory process either comes from other modules, from the E-node, or from receptor nodes, which transduce physical stimulation (e.g. light, sound, displacement, etc.) into activations. Initially, when nothing has been learned yet, all modifiable connections leading to the R-nodes will have equal weights (e.g. midway between the minimum and the maximum weight). So, if a stimulus is presented all R-nodes will be activated equally. Each R-node will activate its corresponding V-node (see Figure 2.3), which will cause competition due to the mutual inhibitory connections of the V-nodes. The R-nodes also excite the A-node, and in this way feed

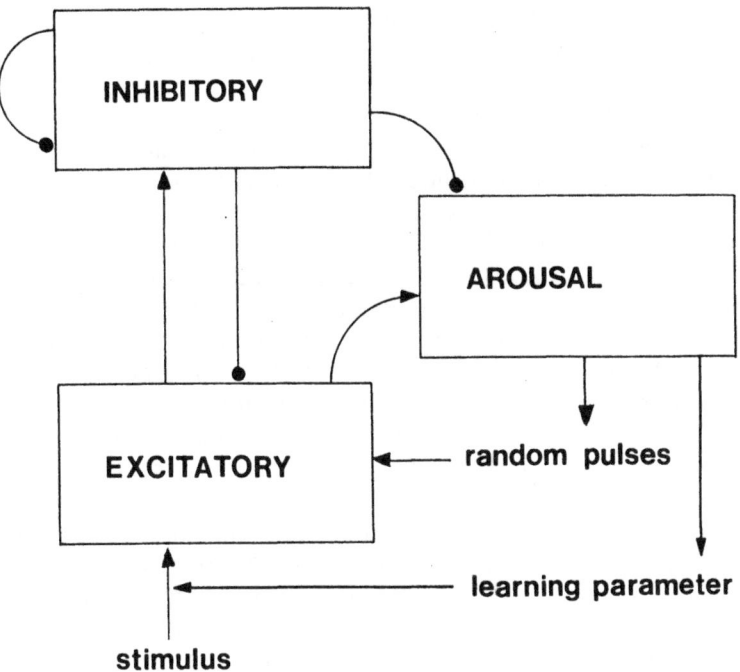

Figure 2.5 Three principal processes in the CALM module. Reprinted with permission from *Neural Networks*, vol. 5, Murre, J.M.J., Phaf, R.H. and Wolters, G., CALM: Categorizing and learning module, Copyright 1992, Pergamon Press Ltd.

the arousal mechanism. In short, the excitatory mechanism's main function is to activate the two other systems.

The inhibitory process, which consists of the activation of the V-nodes, is driven only by the excitatory system. As soon as a V-node becomes activated it starts to inhibit the A-node, the R-nodes, as well as the other V-nodes. The mutual inhibition of the V-nodes causes competition between these nodes. The V-nodes will perform slow oscillations as long as the competition continues. In fact, in a discrete representation of time such oscillations may be seen as an indicator of competition (e.g. Phaf *et al.*, 1990b). Activation of the inhibitory process is, thus, in most cases signalled by the occurrence of oscillations. With regard to the inhibition of the R-nodes by the V-node, a distinction must be made between the inhibition of the matched R-node and the inhibition of other, non-matched R-nodes. The latter inhibition contributes to the competition by strongly undermining the activation sources of its competitor V-nodes. This is essentially a mechanism of recurrent lateral inhibition, which depends on the specific values of the cross weights. The greater the value of the (negative) cross weights the stronger the veto effect. Together with the cross weights the flat weights regulate the fierceness of the competition and the rate of its resolution. When one R-node continually receives just a little bit more input than the others, then after a few iterations the competition will usually be resolved with the present parameter values (see Appendix A). The inhibition to the matched R-node will be smaller than the inhibition to the other R-nodes. The down weight will usually have only a small value and will limit the activation value of the winning R-node. A single CALM module or a CALM network with only feedforward connections will also function well with zero-weight values for this connection. With bidirectional connections between modules, however, small negative values for the down weights will prevent the occurrence of strong 'reverberatory' loops between two winning R-nodes in two different modules. Without down weights the activations of the R-nodes would not decay after the presentation of the input pattern has been stopped. Moreover, the down weight can be seen as a differentiating connection that makes it easier for new patterns to override old activations.

The arousal process, involving the A-node and the E-node, has two important functions. The first is to make possible the resolution of competition when several R-nodes receive the same amount of excitation and the competition cannot be decided. By sending random activations to the R-nodes the balance is disturbed, so that one of the R–V-node pairs is able to win the competition. When the competition has been solved, the random pulses will contribute to the activation of the winning node and enhance the contrast between winning and losing activations. The second function of the arousal system is to control learning rate. As described above, the learning rate is proportional to the E-node activation. When much learning is required, i.e. in novel situations, the connection weights will change quickly. Familiar representations, however, will only be strengthened further; they do not need a sudden reorganization of weights to the winning node. The general purpose of deliberately introducing noise, especially in the learning phase when new patterns are presented and learned, is to create stable representations that are resistant to noise. Only those representations that are not easily disturbed by small variations in input will eventually reach an activation learning phase with every presentation. The module can be said to continue searching for a suitable representation until stable, noise-resistant representations have been found.

The presence of arousal activations is completely determined by the wiring of the module. The ratio high/low weight is chosen in such a way that, if a single R–V-node pair is activated, the total input to the A-node is negative and the A-node activation is suppressed. If, however, more than one pair is activated at the same time, the activation of the A-node will be raised. This is a consequence of the mutual inhibition, which is present only between V-nodes and not between R-nodes. When more then one R–V pair is activated, the excitation of the A-node by the R-nodes will be higher (in absolute magnitude) than the inhibition by the V-nodes. So, a net excitation of the A-node will result, which signals the presentation of a novel, not yet learned, pattern. When a novel input activation pattern is presented to the module (assuming that before presentation the activations were zero or close to zero), a wave of

activations will reach several R-nodes. This causes high initial activation of the
A-node (and subsequently of the E-node). Due to the competition between
V-nodes, the activations will oscillate until the competition has been resolved.
During the oscillations the amplitude of the winning R–V pair gradually
increases. Finally, the oscillation will cease and the winning pair will reach
some stationary level, while all other R–V pairs have zero activation. We shall
call the process of resolving competition 'convergence' or 'relaxation'. The
winning node will maintain some activation as long as the input pattern is
present. If the input vanishes, the activations in the module will decay towards
zero, but the modified interweight will ensure the long-term storage of the
pattern. Such a storage is evidenced by a repeated presentation of the same
pattern. The excitatory process will again be involved by the presentation of
the old pattern. The inhibitory and the arousal processes, however, will be
activated to a lesser degree, because the old R-node will now have a considerable advantage over the other R-nodes. There will not be enough competition
to start a search for a new representation and the old representation will only
be strengthened further. So, the interaction of excitatory, inhibitory and
arousal processes can only give rise to a new representation when such a
representation is needed. The representation will be chosen from the nodes
that take part in the competition. Nodes that have already been committed
will, generally, not take part in the competition. Old patterns will, therefore,
lead to the direct activation of the old representation, whereas new patterns
result in an elaborative search among nodes that mostly have not yet been
committed to other representations. The interplay of these processes will help
to reduce interference at a modular level.

2.2.2 Categorization and learning

The single most important feature of a CALM module is its ability to categorize input activation patterns autonomously. It enables learning without supervision at every module of the network. The local input pattern of a module
is 'chunked' into a superordinate category. The concept of categorization as
a fundamental process in human memory has a long history in psychology (for
a review, see Wickelgren, 1981) and can be seen to supplement the strictly
associative accounts, which have an even longer history (e.g. see Anderson and
Bower, 1974) and may be traced back to antiquity. We feel that incorporation
of such psychological concepts into neural networks may help to mitigate practical problems with current associative models. An important example of such
a problem is the catastrophic interference in associative networks reported by
McCloskey and Cohen (1989) and Ratcliff (1990). The occurrence of such
interference has, in fact, been predicted by Wickelgren (1981, p. 31):

> we need chunking the most to avoid the enormous associative interference
> problem that would result from associating concepts only by associating their

featural constituents, each of which participates in thousands of other concepts and so would have thousands of competing associations.

We shall not discuss the advantages of categorizing networks above strictly associative networks in any detail here, but continue with an illustration of the categorization process in a CALM module (see Chapter 7 for an extensive review).

Categorization in the CALM module is operationalized as the association of a certain input pattern with a unique R-node that is then said to represent the pattern. During and following the categorization process learning takes place, which preserves the association between pattern and R-node by adjusting the interweights to the R-node. In Figure 2.6 different stages of the categorization process have been shown for the module in its simplest form. Of the connections only the interweights and the AE weight are shown. The module in Figure 2.6 receives input from two nodes (e.g. R-nodes in some other module) of which only the first one is activated, with some constant activation value.

In Figure 2.6(a) the module is in its initial state. All nodes have zero activation values. The interweights all have the same initial value.

In 2.6(b) the stimulus has started the excitatory process, while the arousal and the inhibitory process are not yet activated. Note that initially both R-nodes have exactly the same activation value, because they receive the same input.

In 2.6(c) the inhibitory and arousal process have also been activated. Both V-nodes are activated equally. In the next two iterations (which have not been shown) the E-node sends random activation pulses in such a way that, by chance, the second R-node receives more activation than the first one. Consequently, in 2.6(d) the second V-node is (slightly) more activated than the first one. In 2.6(e) the effect of the higher activated V-node can be seen: the left R–V pair has hardly any activation left. Finally, in 2.6(f–h) the module converges to a stationary activation configuration.

During and following the convergence process learning takes place. As was observed earlier, novel stimuli lead to elaboration learning, which has as important characteristics (i) that learning is enhanced and (ii) that random activation is distributed over the R-nodes. In 2.6(b–f) it can be seen how both R-nodes become activated and remain in oscillation, until the E-node noise has resolved the deadlock situation. In 2.6(b–c) all weights from the activated input node increase by the same amount, so that, initially, all possible representations for this input pattern will be changed somewhat. The weights from the input node that has not been activated will decrease strongly. In 2.6(d–g) the first R-node gradually reaches zero activation, so the weights to this node will no longer change. Both weights to the second node change considerably. In the meantime, the A-node receives excitation from the R-nodes and inhibition from the V-nodes. Both the V-nodes and R-nodes oscillate, but on average the inhibition from the V-nodes will be lower in absolute magnitude than the excitation from the R-nodes because of the mutual inhibition of

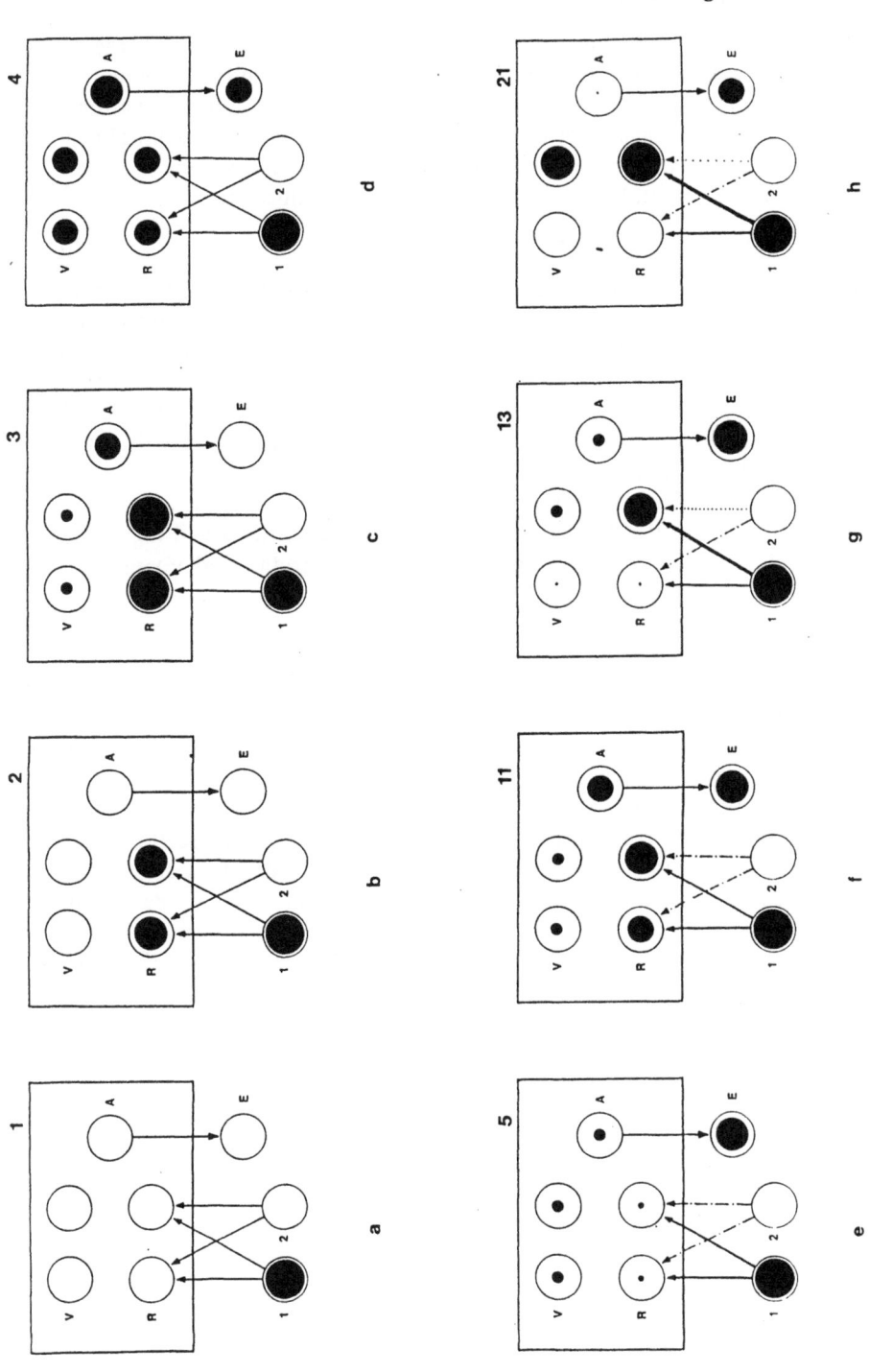

Figure 2.6 Snapshots at iterations 1, 2, 3, 4, 5, 11, 13 and 21 illustrating categorization and learning in a two R-node CALM module. The size of the full circle in every node indicates its activation level, and the thickness of a line indicates the magnitude of the weight on the connection. For a detailed explanation, see text. Reprinted with permission from *Neural Networks*, vol. 5, Murre, J.M.J., Phaf, R.H. and Wolters, G., CALM: Categorizing and learning module, Copyright 1992, Pergamon Press Ltd.

the V-nodes. As a result, the A-node will be activated during the convergence process. On average, the E-node activation will be higher and less fluctuating than the A-node activation, because it integrates over the A-node activations.

The left weight increases, owing to some form of 'Hebbian learning' (i.e. the positive term in the learning rule is non-zero). The right weight decreases because, since $a_j = 0$ and $\Sigma a_f = 1$, the learning rule reduces to

$$\Delta w_{ij} = - \mu_t a_i L w_{ij} w_{i1} a_1 < 0 \qquad (2.5)$$

In the example, the input pattern is encoded in the second R-node. At that moment of convergence the E-node is still highly activated. This means that learning is enhanced and that random noise is being distributed. The increased learning ensures that the pattern becomes encoded quickly. The noise further contributes to this effect because it increases the activation of the second R-node; in the other R-node the added noisy activation is suppressed by the recurrent inhibition from the highly activated second V-node. Following the decaying E-node (past iteration 21, see Figure 2.6(h)) learning will gradually slow down until it settles at its base rate. Elaboration learning will then have shifted back to activation learning with base rate learning and no random activations.

With the next presentation of the same input pattern (1,0), the second R-node will have a direct advantage over the other, because the weights to the winning R-node have increased, while the weights from this pattern to the other R-node have decreased. Though the A- and E-node will become activated to some degree, the elaboration phase will be much shorter and the module will almost directly shift towards activation learning. In this way, repeated presentation of a pattern leads to the same categorization, and this categorization will be reached much faster as will be shown in Section 3.2.

In summary, repeated presentation of a pattern will lead to a shift from elaboration learning towards activation learning. This shift is accompanied by a reduction in convergence time. *Both the effects of repeated and prolonged presentation of a pattern* may be described as forms of habituation, the habituating factor in this case being the response of the arousal system.

2.2.3 Separation of correlated patterns

As has been discussed by Carpenter and Grossberg (1987), the learning rule used in CALM has the important property of allowing discrimination of correlated or non-orthogonal patterns, such as (1,0) and (1,1). Such patterns, when considered as activation vectors, can be defined as having a non-zero scalar product. By taking the background activation into account the rule manages to discriminate between non-orthogonal patterns. The operation of the learning rule with the presentation of the two correlated patterns is shown in Figure 2.7. In the figure the patterns (1,0) and (1,1) are presented to a two R-node CALM module. Suppose that (1,0) becomes categorized at the left

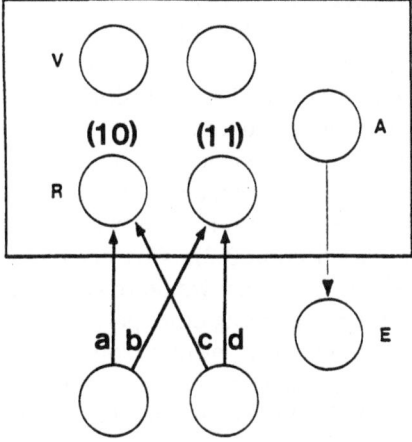

Figure 2.7 Relative sizes of weights a, b, c and d must satisfy the conditions $a > b$ and $b + d > a + c$ in order to discriminate between the patterns (1 0) and (1 1). Reprinted with permission from *Neural Networks*, vol. 5, Murre, J.M.J., Phaf, R.H. and Wolters, G., CALM: Categorizing and learning module, Copyright 1992, Pergamon Press Ltd.

R-node, and (1,1) on the right. In order to obtain such a categorization it is easily seen that the weights a, b, c and d must satisfy the following conditions:

$$a > b \quad \text{and} \quad b + d > a + c$$

Only with this weight configuration will the left R-node represent pattern (1,0) and the right node (1,1). How does this configuration come into being through the mere presentation of the two patterns? Initially, the weights are taken to be equal and about half-way between their minimum and maximum value. When pattern (1,0) is presented, weight a will increase, because it has no neighbour activations. Weight c, however, carries no activation and has one activated neighbour connection. This neighbour activation causes a decrease of weight c. When pattern (1,1) is presented, weights b and d will be increased equally. Due to the competitive character (i.e. competition between the weights) of the Grossberg learning rule this increase will be smaller than when only a single input node is activated.

With a proper choice of the parameters K and L both conditions will then be preserved. It should be noted that with lower initial values of the inter-weights the correlated patterns will not be automatically separated. Separation in a winner-take-all framework requires that with presentation of different patterns different R-nodes are maximally activated. If initial weights are too low, it is possible that the second pattern presented cannot be classified on a different node, because the weights to this node can be lower than to the previous winner (see the analysis in the next section).

A further observation should be made about the dependence of separation upon the difference between the two patterns. It is evident that when the patterns become more similar they will not be able to cause significant differences in maximal activation of the R-nodes. There should, therefore, be some point where separation breaks down. This transition is not discontinuous but gradual and forms a complicated function of the arousal process and the already stored patterns. In fact, when all R-nodes have been committed and discrimination can no longer proceed, a new pattern is classified on the R-node with the most similar representation. In other words, some form of generalization takes place (see also Section 3.3). This behaviour differs from ART 1 (Carpenter and Grossberg, 1987) that would reject dissimilar patterns when all nodes have been used for representing other patterns. In the case of a not completely 'filled' CALM module the transition from discrimination to generalization is determined by the parameters as it is in ART. Although the nature of the search process in CALM differs considerably from that in ART, the ratio of low and high weight may have a similar effect on the arousal system as ART's vigilance parameter. The actual discrimination and generalization behaviour of CALM will be demonstrated in the next chapter, where simulations with single modules of various sizes will be presented.

2.2.4 Equilibrium weight values

Because weights in CALM are bounded by the learning to values between 0 and K, one might wonder what the effect of prolonged learning is. If a given pattern is presented to the CALM module for a very long time, will the weights finally reach extreme values 0 or K, or will they equilibrate at some value intermediate between 0 and K? We may also ask: in what way do such equilibrium values depend on the activation values in the input pattern?

Carpenter and Grossberg (1987, their Appendix) show that the learning rule by Grossberg (i.e. in its original form) obeys what they call the 'Weber law'. With this, they mean that the equilibrium weight values, w, are inversely related to the number of active elements in the input pattern, n:

$$w = \frac{A}{B + n} \qquad (2.6)$$

with suitably chosen values for the constants A and B. Though the CALM version of Grossberg's learning rule is somewhat more complex, it is still possible to derive equilibrium values. These values do not follow a straightforward Weber rule. In the limit (i.e. for large n) they follow a rule of the form

$$w = \sqrt{\frac{K}{Ln}} \qquad (2.7)$$

where K and L are constants from the learning rule as used throughout this book. The result can be found by equating the learning rule (equation (2.2)) with zero (i.e. no more weight changes: the weight values are at equilibrium):

$$\Delta w_j = \mu_t a_i \left((K - w_j)a_j - L w_j \sum_{f \neq j} w_f a_f \right) = 0 \qquad (2.8)$$

where w_j is the weight from node j to node i (i.e. the subscript i is suppressed). Suppose that node j has n activated neighbours (as defined above, e.g. in Figure 2.4), and that these n neighbour activations are all equal to some value $a = a_j$. Suppose, furthermore, that initially all weights are equal (as we have always chosen in our simulations), say with value w. Then, we have

$$\Delta w = \mu_t a_i [(K - w)a - Lwnwa] = 0 \qquad (2.9)$$

or

$$K - w - Lnw^2 = 0 \qquad (2.10)$$

from which we derive (for values for K and L as used throughout this book, see Appendix A)

$$w = \frac{\sqrt{1 + 4KLn} - 1}{2Ln} \qquad (2.11)$$

For example, with two activated input nodes (n is 1, i.e. each node has exactly one neighbour), we find an equilibrium value of 0.618, which, incidentally, is a well-known proportion in art and architecture. With five nodes ($n = 4$; four neighbours), we find 0.390. Note that the equilibrium weight values are not dependent on the absolute magnitude of the input activations, nor on the magnitude of the post-synaptic activation. They only depend on the *number* of activated nodes. By taking n to be very large the reader can easily derive the (limit) formula given above (equation (2.7)).

Contrary to ART 1, CALM uses continuous input activations, so that we must also investigate equilibrium weights in case we have input activation values intermediate between 0 and 1. In general, this is more complicated to analyse, but we may consider the situation where we have only two activated input nodes. Suppose that two nodes, node 1 and node 2, are connected to some activated R-node, with weights called w_1 and w_2, respectively. Suppose, furthermore, that they have constant and possibly different (real-valued) activations a_1 and a_2, respectively. As in the analysis above we set the change in weight equal to zero. From this, similar to equation (2.10), we now derive two equations to be solved simultaneously:

$$(K - w_1)a_1 - L w_1 w_2 a_2 = 0 \qquad (2.12)$$

and

$$(K - w_2)a_2 - L w_1 w_2 a_1 = 0 \qquad (2.13)$$

From equation (2.12) we derive

$$w_2 = \frac{(K - w_1)a_1}{L w_1 a_2} \qquad (2.14)$$

and, from equation (2.13)

$$w_1 = \frac{(K - w_2)a_2}{Lw_2a_1} \qquad (2.15)$$

To find the relation between input activations and equilibrium weights, we further investigate the ratio w_2/w_1:

$$\frac{w_2}{w_1} = \frac{(K - w_1)a_1 Lw_2 a_1}{(K - w_2)a_2 Lw_1 a_2} \qquad (2.16)$$

After some algebraic manipulation, we find

$$\frac{K - w_2}{K - w_1} = 1 + \frac{w_1 - w_2}{K - w_1} = \frac{a_1^2}{a_2^2} \qquad (2.17)$$

and

$$w_1 - w_2 = \left(\frac{a_1^2}{a_2^2} - 1\right)(K - w_1) \qquad (2.18)$$

from which we derive the relation between w_2 and w_1:

$$w_2 = \left(\frac{a_1^2}{a_2^2}\right)w_1 + \left(1 - \frac{a_1^2}{a_2^2}\right)K \qquad (2.19)$$

In other words, the relation between w_1 and w_2 is dependent only on the *ratio* of a_1 and a_2, and not on the absolute magnitude of a_1 and a_2. For $K = 1$, the maximum weight value used primarily throughout this book, we may write

$$w_2 = 1 - A^2(1 - w_1) \quad \text{with} \quad A = a_1/a_2 \qquad (2.20)$$

For the case $a_1 = a_2$, we have $w_1 = w_2$. The equilibrium values for this case were analysed above and are given by equation (2.11).

These results remain valid, with some slight alterations, if at equilibrium non-zero neighbour activations (with non-zero weights) are present, say for a_3 to a_n. To analyse this case we extend equations (2.12) and (2.14) with a term E to take into account the influence of neighbour activation:

$$(K - w_1)a_1 - Lw_1(w_2a_2 + E) = 0 \quad \text{with} \quad E = \sum_{f=3}^{n} w_f a_f \qquad (2.21)$$

From this we derive for w_2

$$w_2 = \frac{(K - w_1)a_1 - Lw_1 E}{Lw_1 a_2} \qquad (2.22)$$

As above we write the ratio

$$\frac{w_2}{w_1} = \frac{[(K - w_1)a_1 - Lw_1 E]\,Lw_2 a_1}{[(K - w_2)a_2 - Lw_2 E]\,Lw_1 a_2} \qquad (2.23)$$

$$\frac{w_2}{w_1} = \frac{(K - w_1)a_1^2 w_2 - w_1 w_2 E a_1}{(K - w_2)a_2^2 w_1 - w_1 w_2 E a_2} \tag{2.24}$$

$$\frac{Ka_1^2 - w_1(a_1^2 + E a_1)}{Ka_2^2 - w_2(a_2^2 + E a_2)} = 1 \tag{2.25}$$

$$\frac{K - w_2(1 + E a_2^{-1})}{K - w_1(1 + E a_1^{-1})} = \frac{a_1^2}{a_2^2} \tag{2.26}$$

From this, similar to the derivation of equation (2.19), we find

$$w_2 = w_1 \left(\frac{a_1^2}{a_2^2}\right)\left(\frac{1 + E a_1^{-1}}{1 + E a_2^{-1}}\right) + \frac{[1 - (a_1^2/a_2^2)]}{(1 + E a_2^{-1})} K \tag{2.27}$$

Combining the above results we can gain some understanding of the weight values emerging during the learning process. For example, if we have two activations $a_1 = 0.4$ and $a_2 = 0.2$, we find equilibrium values of $w_1 = 0.843$ and $w_2 = 0.372$ in the simulations. These remain unchanged if we multiply a_1 and a_2 by some constant, so that, for example, $a_1 = 1.0$ and $a_2 = 0.5$ give the same equilibrium weight values as $a_1 = 0.4$ and $a_2 = 0.2$, which is the result predicted by equation (2.19). If we have five nodes with an activation of 1.0 each, after many iterations we find equilibrium weights of 0.390 in the simulations. These results are exactly as predicted by equation (2.11) (with $n = 4$). Now, if we run a combined simulation with $a_1 = 0.4$, $a_2 = 0.2$ and $a_3 = 1.0, ..., a_7 = 1.0$, we find equilibrium weights $w_1 = 0.172$, $w_2 = 0.092$ and $w_3 = 0.382, ..., w_7 = 0.382$. These results are close to the values predicted by equations (2.11) and (2.27), as the reader can easily verify. For example, in equation (2.27) we have neighbour activation $E = 0.382 \times 5 = 1.91$. Substituting $w_1 = 0.172$ in equation (2.27) yields $w_2 = 0.090$, where the slight difference with the simulation result is probably due to approximation errors.

Summarizing, we see that the emerging weights are the result of a complex interplay of the pre-synaptic activations. Weight values tend to reflect activation values, whereby the *ratio* of two weights is primarily dependent on the *ratio* of the activations and, to a lesser extent, on the relative magnitude of the overall neighbour excitation. The absolute magnitude of the weight values is determined by the relative neighbour excitations. In case all activated nodes have the same activation value, it is only the *number* of activated nodes that determines the equilibrium weight values.

Chapter 3

Simulation Studies of Performance and Self-organization in CALM

3.1 Simulation with CALM

Although it is perfectly possible to understand the functioning of CALM in a global manner by speculating about its potential behaviour, a more precise verification is needed. In CALM, because the activation rule and learning rule are non-linear, and because of the complex interaction of the three main processes, a rigorous mathematical analysis of the behaviour of a single CALM module appears to be very difficult. Although certain aspects of CALM are amenable to analysis (e.g. see Section 2.2.4), to investigate its behaviour we have to take recourse to simulations with single CALM modules. In this chapter, some simulation studies will be presented concerning the dependence of convergence time on module size, of convergence time on pattern repetition, and of discrimination time on problem size with orthogonal patterns. Discrimination and generalization of various pattern sets with non-orthogonal patterns will also be demonstrated. The simulation studies are followed by an illustration, a small CALM model that learns the EXOR relation. The final section presents an extension of the CALM module to a variant module, called CALSOM, that is able to self-organize its representation in accordance with the topology of the input set. The variant module is derived with only minor changes in the original CALM module.

Unless mentioned otherwise, all simulations described were carried out with the same set of model parameters. This set is listed in Appendix A, together with some comments on how we have obtained these values. Furthermore, it must be stressed that, unless mentioned otherwise, learning was never 'switched off'. In CALM there is no need for separate training and testing phases, and no desired output needs to be specified. Learning in the simulations in this chapter always proceeded without supervision or explicit instruction.

3.2 Convergence time and discrimination time

To study the influence of module size on convergence time we recorded convergence times when presenting patterns of the type $(0,0, ..., 1,0, ...)$ to different modules, that had not learned anything else before (and were thus in their initial state). Convergence time refers to the time (i.e. number of iterations) required by the module to categorize a pattern. The criterion for convergence was that one of the R-nodes reaches some threshold c_h, while all others are below some threshold c_l (see Appendix A). The exact values for these thresholds are not critical. Changes will have only slight and consistent

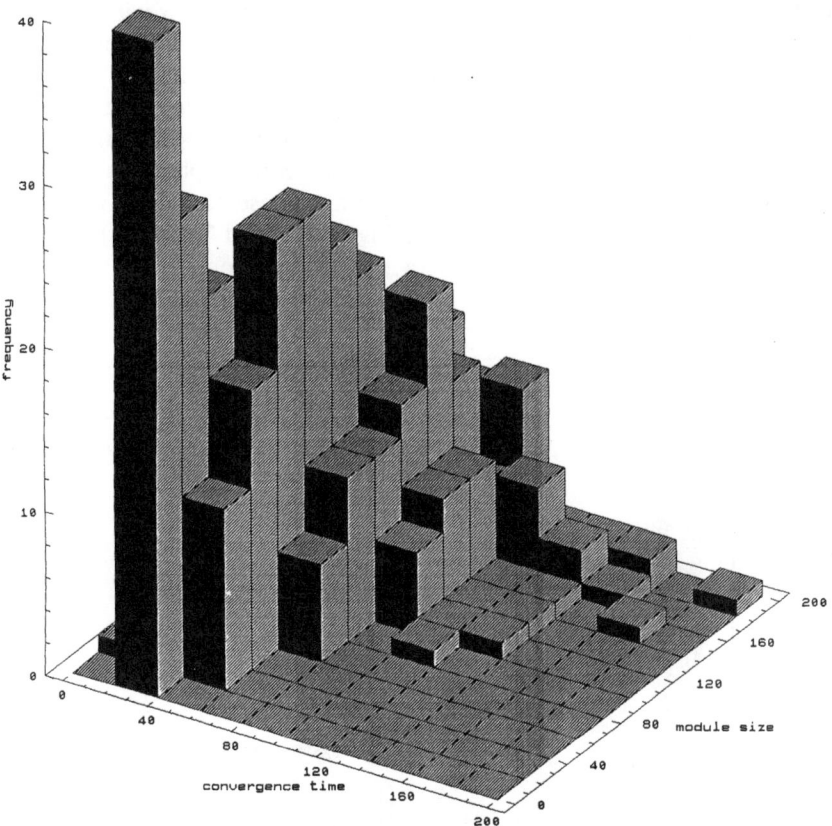

Figure 3.1 Frequency diagram of module size versus convergence time. Convergence time is shown in intervals of 20 iterations. For every module size interval of 10 the pattern was categorized 20 times while starting from the same initial state. In every bar of the histogram the 20 presentations to each of two modules of subsequent sizes are combined. The module size dimension thus represents modules sizes $10 + 20$, $30 + 40$, ..., $190 + 200$, and the frequency dimension has 40 (i.e. 2×20) as its maximum value. Reprinted with permission from *Neural Networks*, vol. 5, Murre, J.M.J., Phaf, R.H. and Wolters, G., CALM: Categorizing and learning module, Copyright 1992, Pergamon Press Ltd.

influences on all convergence times. Only if the difference between c_h and c_l is too small will 'false' convergences occur, but at the values chosen in the simulations there is little danger of this. Although the patterns used were very simple, this categorization task is not particularly easy, because the module has to decide upon a suitable uncommitted node. Initially, learning weights are all equal and hence the set of uncommitted nodes is of maximum size. Convergence, therefore, usually takes longest with presentation of the first pattern (see also the next section). Module sizes of 10, 20, 30, ..., up to 200 R-nodes were used. Because the behaviour of the module is stochastic, the same pattern was repeated 20 times for every size, each time starting anew with the original (equal) initial weights. The frequency of convergence as a function of convergence time and module size is plotted in the histogram of Figure 3.1. Both convergence time and the variance in convergence time clearly increase with module size. In the figure it is easily verified that, for modules up to size 60, convergence on the first presentation of a pattern takes place within 60 iterations (the average convergence time at the first presentation for a module of size 60 is about 33 iterations; see also Figure 3.2 and Table 3.1). For larger modules convergence usually occurs within 80 iterations.

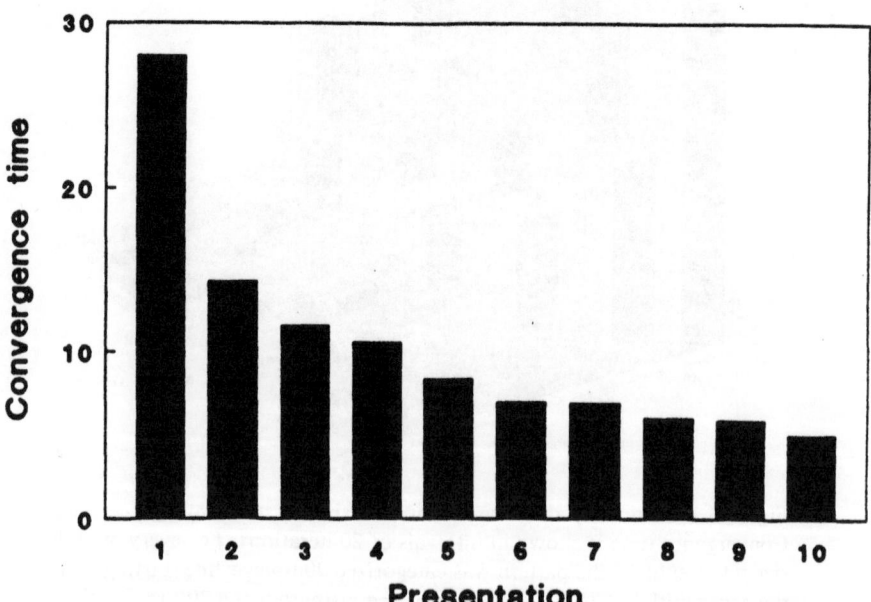

Figure 3.2 Average convergence time as a function of pattern repetition. The figure illustrates decreasing convergence time with repetition of a pattern. The size of the module was 20. The data are averaged over the 20 patterns in a pattern set. Reprinted with permission from *Neural Networks*, vol. 5, Murre, J.M.J., Phaf, R.H. and Wolters, G., CALM: Categorizing and learning module, Copyright 1992, Pergamon Press Ltd.

Table 3.1 Average convergence times (number of iterations) for modules of size 20, 40, 60, 80 and 100 at first to tenth presentation of an orthogonal pattern set. Averages are taken over the entire pattern set.

Size	Presentation									
	1	2	3	4	5	6	7	8	9	10
20	28.0	14.3	11.6	10.6	8.4	7.0	6.9	6.0	5.9	5.0
40	28.3	14.3	10.9	13.3	13.1	11.9	10.4	9.4	8.9	8.3
60	32.5	15.8	11.9	10.4	12.4	12.9	11.9	10.6	10.1	9.2
80	36.7	17.1	12.6	11.6	10.9	12.5	12.5	11.7	11.4	10.3
100	36.6	17.4	13.4	11.8	10.4	11.4	12.0	12.5	12.1	11.9

Patterns that have been presented before give rise to faster convergences, because the increased weights to the R-node representing a repeated pattern give it a distinctive advantage. With every presentation this advantage becomes larger due to further learning. Moreover, learning gradually shifts from elaboration to activation learning. Figure 3.2 illustrates this by showing that the repetition of patterns leads to a marked decrease in convergence time. In this simulation a set of 20 orthogonal patterns of the form (0, ..., 0, 1, 0, ..., 0) was presented to a CALM module of size 20. In these patterns only one activation is equal to one; all others are zero. As in all simulations in this work initial interweights were all equal. Each pattern was presented 10 times without initializing the weights in between presentations (i.e. the entire pattern set was presented 10 times as a fixed sequence of patterns). In the figure the average convergence time of the set is plotted as a function of subsequent presentations. The convergence time drops 50% at the second presentation, for reasons mentioned in the previous section. Following the second presentation convergence time continues to decline at a lower rate. With orthogonal pattern sets ultimately convergence times of one iteration can be reached, because after prolonged learning every input node will have exactly one strong connection to an R-node and zero-weight connections to all other R-nodes. If that is the case, the convergence criterion is met at the first iteration. With non-orthogonal patterns, however, there will always exist nodes that have (moderately) strong connections to several R-nodes. The presentation of patterns including such nodes will cause competition, resulting in a convergence time greater than one. The increase in convergence time with module size and the decrease with repeated presentations is also found in the following simulation. Table 3.1 shows average convergence times for modules of sizes 20, 40, 60, 80 and 100 at the first to the tenth presentation of orthogonal pattern sets. (The data for module size 20 are the same as in Figure 3.2.) The pattern sets were as in the previous simulation, but with the number of patterns in the set equal to module size. The steep drop in convergence time by about a half following the first presentation is found for all module sizes. Because of the stochastic character of the categorization process convergence times do not all show a strict decline during the first few presentations. If patterns have been presented more often, however, the contribution of the noise becomes progressively smaller. At later presentations a more regular curve can be observed.

A CALM module is in principle capable of categorizing an arbitrary input pattern. To investigate how problem size (i.e. number of learned patterns) affects total discrimination time (time until 95% correct discrimination of all patterns is reached), we prepared sets of orthogonal patterns of the type $(0, ..., 0, 1, 0, ..., 0)$. The pattern sets were presented to modules of sizes 5, 10, 15, ..., up to 95. The number of patterns always matched the size of the module. After a pattern had reached the convergence criteria, it was presented for 100 additional iterations to stabilize learning completely. The presentation of a pattern set (without resetting the weights) was repeated until 95% correct discrimination had been reached, e.g. in modules of sizes 20 to 35 no more than one pair and in modules of sizes 40 to 55 no more than two pairs of patterns were categorized on the same node. In the latter example this corresponds to maximally two uncommitted nodes in a module. In Figure 3.3 total discrimination time t_d (total number of presentations = number of patterns × repetitions until 95% discrimination) is plotted against problem size N (number of patterns in the set). A multiplicative regression analysis yielded the following polynomial relation for the scaling behaviour of the module:

$$t_d = aN^b \qquad (3.1)$$

with $a = 0.2$ and $b = 1.7$. For similar tasks using an error backpropagation algorithm, values for b in the order of 3.5 have been reported (Morse, 1989). So, the discrimination of a complete set of orthogonal patterns as a function of problem size in CALM scales favourably compared with systems using error backpropagation. Of course, the relation found for CALM refers to unsupervised learning whereas the backpropagation exponent refers to supervised learning. Therefore, the learning speeds may not be compared directly. It can be argued, however, that unsupervised categorization is more difficult than categorization with explicit instruction. One of the reasons that backpropagation takes longer and requires supervision is that different patterns are superimposed on the same set of weights. In CALM the patterns will be encoded as distinctly as possible. It is evident that this not only enables CALM to categorize without supervision but also that it will reduce interference, because the same connections no longer have to be used to store different patterns (see also the discussion in Chapter 7). It may be noted that in CALM also supervised learning is possible by preactivating the desired R-node. Although in this case discrimination will be perfect and becomes almost trivial, paradoxically, learning will proceed slower because there will be less competition which leads to a decreased learning rate. With an optimal choice of parameter values, however, performance will still compare favourably with backpropagation.

In real-world problems, generally patterns will not be orthogonal, but show complicated relationships. In some cases the differences between patterns have to be ignored, whereas in other cases small differences are critically important.

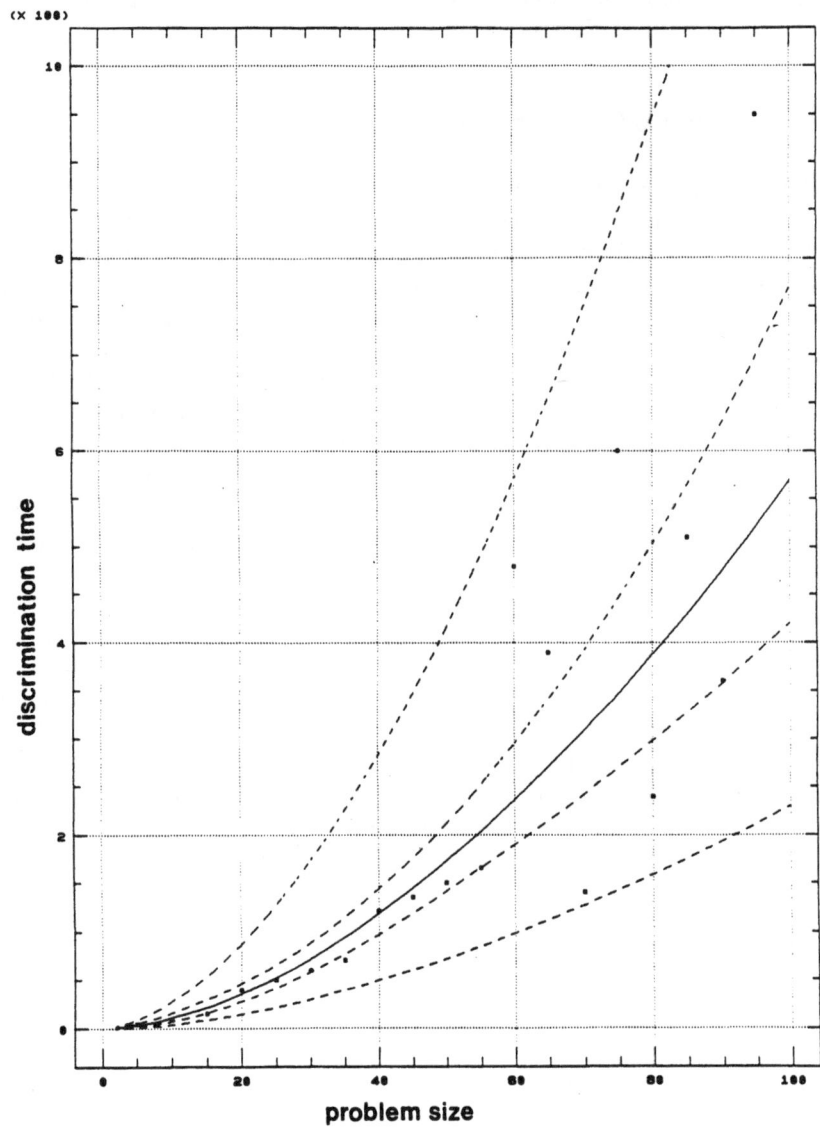

Figure 3.3 Average time (in number of presentations) until 95% correct discrimination as a function of problem size. A polynomial function has been fitted, together with 5% and 10% confidence intervals. Reprinted with permission from *Neural Networks*, vol. 5, Murre, J.M.J., Phaf, R.H. and Wolters, G., CALM: Categorizing and learning module, Copyright 1992, Pergamon Press Ltd.

In the next section it will be demonstrated to what extent non-orthogonal patterns are discriminated and how a CALM module may generalize over patterns which form natural clusters.

3.3 Discrimination and clustering

When a small CALM module of size 3 is presented with three patterns, (0,1), (1,0) and (1,1), it will categorize each of these on a separate R-node with little difficulty. If the difference between two patterns becomes too small, however, the CALM module tends to categorize both on the same R-node. To illustrate this, two patterns, (1.0,0) and (1.0,x), where x was varied between 0.1 and 1.0 in steps of 0.1, were presented to a CALM module of size 2. Both patterns were presented 10 times for 50 iterations, without initializing the learning weights in between presentations (but activations were initialized). This procedure was replicated 10 times for each of the 10 values for x. It was found that, if the Euclidean distance between the two patterns exceeded 0.7, the module consistently learned to discriminate between the two patterns. If the Euclidean distance was less than 0.6 both patterns were consistently categorized on one R-node, leaving the other node uncommitted. For a distance of 0.6 the module learned to discriminate between the two patterns in two out of 10 presentations, for a distance of 0.7 in seven out of 10 presentations.

The distance between two patterns is not the only factor that determines whether they will be categorized on the same node or on separate nodes. The size of the CALM module is an important factor as well. Large modules can more easily accommodate patterns than small modules. If the number of patterns is large relative to the size of a module, it is forced to combine several patterns on to one representation. In other words, it will generalize over patterns. To study the generalizing capacity of the CALM module eight patterns were prepared as shown in Table 3.2(a). Table 3.2(b) represents the matrix of squared Euclidean distances between the patterns. As can be seen, the patterns form two principal clusters. These clusters are defined by the fact that the distance between any two patterns in the same cluster is smaller than the distance between any two patterns from different clusters. The patterns were presented to CALM modules of increasing sizes of 2, 3, ..., up to 8 R-nodes. The set of eight patterns was presented in the order listed in Table 3.2(a). Each pattern was presented for 40 iterations. The entire set was presented 10 times to every module, at which point a stable categorization for all patterns in the set had usually been reached. The whole procedure was replicated 10 times, so that an indication could be obtained about the consistency with which particular categorizations were established. In Table 3.2(c) the frequency of particular categorization patterns is shown for the 10 replications. The nodes have been given labels a, b, c, ... Thus, if eight patterns were presented to a CALM module of size 2, patterns 1 to 4 were categorized on one node, labelled a, whereas patterns 5 to 8 were categorized on the other node, labelled b.

Table 3.2 Generalization by a CALM module. (a) Eight patterns presented to the module. (b) Squared Euclidean distances between the patterns. (c) Categories formed with modules of sizes 2, 3,..., 8. Letters, a, b,..., are used to label nodes, e.g. the final categorization pattern aaaabbbb means that the first four patterns are categorized on one node and the last four on the other. Frequency refers to the number of replications in which this categorization pattern was obtained.

(a)

No.	Pattern								
1	1	1	1	1	0	0	1	0	0
2	1	1	1	1	0	1	0	0	0
3	1	1	1	0	1	1	0	0	0
4	1	1	0	0	1	1	1	0	0
5	0	0	1	1	1	0	0	1	1
6	0	0	0	1	1	0	1	1	1
7	0	0	0	1	0	1	1	1	1
8	0	0	1	0	0	1	1	1	1

(b) Squared Euclidean distances

1	2	3	4	5	6	7	8	
0	2	4	4	6	6	6	6	1
	0	2	4	6	8	8	6	2
		0	2	6	8	8	6	3
			0	8	6	6	6	4
				0	2	4	4	5
					0	2	4	6
						0	2	7
							0	8

(c) Clustering results

Module size	Patterns								Frequency
	1	2	3	4	5	6	7	8	
2	a	a	a	a	b	b	b	b	10
3	a	a	b	b	c	c	c	c	10
4	a	b	c	c	d	d	d	d	7
	a	a	b	b	c	c	d	d	2
	a	a	b	c	d	d	d	d	1
5	a	b	c	c	d	d	d	d	6
	a	b	c	c	d	d	e	e	1
	a	a	b	b	c	c	d	d	1
	a	b	c	d	e	e	e	e	1
	a	a	b	c	d	d	e	e	1
6	a	b	c	d	e	e	f	f	5
	a	b	c	c	d	d	e	e	4
	a	b	c	c	d	e	f	f	1
7	a	b	c	d	e	f	g	g	6
	a	b	c	c	d	e	f	f	4
8	a	b	c	d	e	f	g	g	4
	a	b	c	c	d	e	f	g	3
	a	b	c	d	e	f	g	h	2
	a	b	c	c	d	e	f	f	1

Abstract labels were used, rather than node numbers, to stress the global characteristics of the categorizations, such as the size of the clusters formed. The frequency information in Table 3.2(c) must be read as follows. When the eight patterns were presented to a CALM module of size 8, for example, it can be seen that in two out of 10 replications perfect discrimination took place. In four cases, patterns 7 and 8 were categorized on a single node, leaving one node in the module uncommitted. In three cases patterns 3 and 4 were clustered together, and in one case both 3 and 4 as well as 7 and 8 were combined, leaving two nodes uncommitted.

The table shows that a CALM module clusters patterns on the basis of shortest distance when insufficient R-nodes are available for representing all patterns, even when the distance between these patterns exceeds the minimal distance for complete separation. A CALM module of size 2 consistently makes two clusters of equal size. The larger modules clearly make finer discriminations between patterns. Due to the state-dependent noise in the CALM module some variation can be observed in the clustering arrived at in different replications. Sometimes patterns are clustered together that have larger distances than patterns that are still separated. For instance, with module size 4 one categorization has been obtained where patterns 3 and 4 (with a squared Euclidean distance of 2) are separated, whereas patterns 5, 6, 7 and 8, with larger distances, have been clustered. As can be seen from the table, such miscategorizations occur only infrequently and do not really disturb the general tendency to cluster patterns on the basis of their distance when insufficient R-nodes are available.

An effect of order of presentation can also be observed. Patterns that were presented first (i.e. patterns 1, 2, 3, ...) are separated better than patterns presented later in the series (i.e. patterns..., 6, 7, 8). The foregoing simulations show that the CALM module first attempts to discriminate between patterns. If this proves to be impossible, because there are too many patterns relative to the number of R-nodes, it will generalize and form clusters. If a novel pattern sufficiently resembles an already learned pattern it will have a greater chance of being categorized on the same node if most other R-nodes have been committed. Generalization in a network consisting of multiple modules, which learns in an unsupervised mode, will thus proceed by introducing 'bottlenecks' in one or all of the parallel streams that process information presented to the network. The important question of how generalization and discrimination depend on network architecture will not be treated in any detail in this work but is illustrative of the manner in which CALM networks can be prestructured to accommodate the problem at hand (see Chapter 7 for a brief discussion of the relation between architecture and generalization).

3.4 Illustration: learning the EXOR

The CALM module is a building block for larger networks, not a network model in itself. A network can be prestructured in terms of layers and modules, representing global relations between input and output, as well as parallel streams of processing in the network. Such architectures may exclude the formation of certain associations, whereas others may be favoured. The modeller, however, is not forced to specify every single connection. Only the global connection scheme between modules needs to be specified. Thus, the modular approach stands between totally unstructured and fully handtailored networks. In a multi-modular network the architecture and characteristics of individual modules will impose boundary conditions on the performance of the network as a whole. If the characteristics of individual modules and the architecture are known, it is possible to derive a rough idea of the characteristics of the network. Although we are not yet able to specify the relation between architecture and learning possibilities fully, we shall illustrate this approach, here, by giving an example of a multi-modular network that learns the EXOR relation.

Learning the EXOR logical function has become almost a benchmark test for learning neural networks. Although CALM was developed primarily for applications relying on unsupervised learning, this simulation illustrates how it performs with supervised learning. We presented four input patterns, $(0.1, 0.1)$, $(0.1, 1)$ $(1, 0.1)$ and $(1, 1)$, to an input module (see Figure 3.4). This two-node module merely serves to hold a fixed activation pattern representing the input. The logical function EXOR takes only zeros and ones as its input. We coded the zero with an activation of 0.1, instead of 0.0, because in this way the logical input $(0\ 0)$ (coded as $(0.1, 0.1)$) still gives rise to some activation in the network. The rest of the network consists of two CALM modules. The upper CALM module represents the output activations; the middle CALM module can be seen as a kind of hidden node layer. The problem with the EXOR function is to separate the patterns $(1,1)$ and $(0.1,0.1)$ from the patterns $(0.1,1)$ and $(1,0.1)$. This requires supervised learning, since when learning is unsupervised the latter two patterns would be discriminated quite easily by the CALM module, but should not be discriminated for the EXOR function. The upper CALM module needs only two R-nodes to represent the two possible truth values. The supervision was applied by preactivating one of these R-nodes. The two CALM modules are connected by bidirectional (but not symmetrical) learning connections. This is not strictly necessary for learning the EXOR, but it is found to improve the discrimination time.

In the training phase one of the R-nodes in the upper module was kept at a constant activation of 0.5. The left node was activated if either $(1, 0.1)$ or $(0.1, 1)$ was presented as an input pattern, the right node with presentation of $(1,1)$ or $(0.1, 0.1)$. During a run, each of the four patterns was presented for 45 iterations. After each run the model was tested by presenting it with only input patterns, whilst recording the output. Because further learning might interfere during the test phase, learning was temporarily disabled (but the noise was not). The entire procedure was replicated 20 times, each time starting with equal initial weights. Of the 20 replications, the EXOR relation was learned 11 times after just one training run; another seven took two runs; the

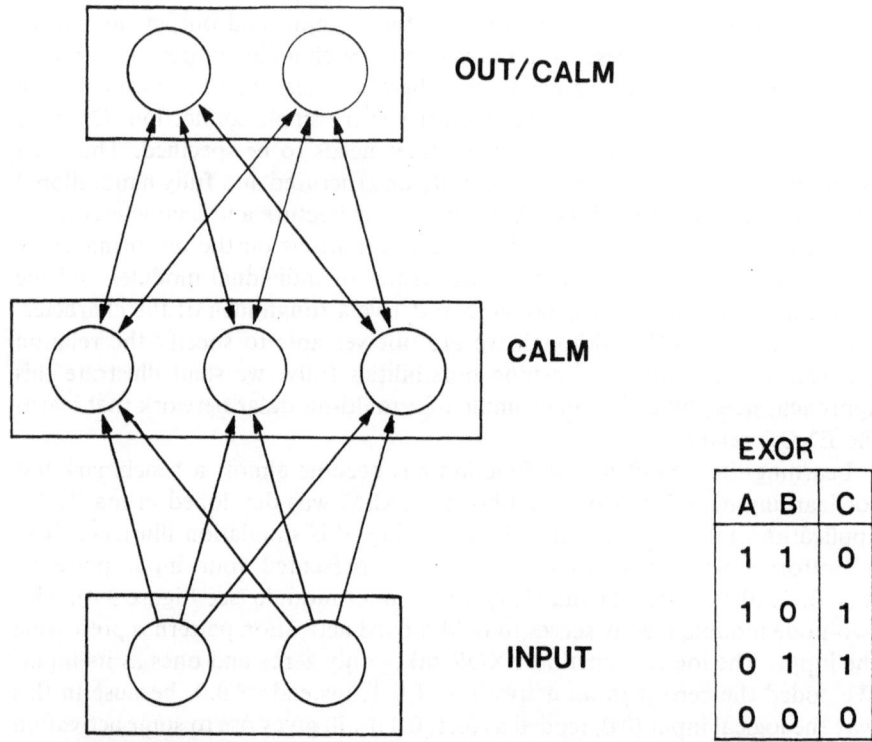

Figure 3.4 Model that learns the EXOR, consisting of two CALM modules and an input module. Activations of nodes in the input module represent the input patterns. The winning activations in the upper CALM module represent the output pattern (0 or 1). The table to the right of the network represents the truth table for the logical EXOR function. Reprinted with permission from *Neural Networks*, vol. 5, Murre, J.M.J., Phaf, R.H. and Wolters, G., CALM: Categorizing and learning module, Copyright 1992, Pergamon Press Ltd.

remaining two replications took three and four runs. On average the EXOR relation was learned in 1.6 runs of 45 iterations per pattern, which compares quite favourably with other learning neural networks such as backpropagation (Rumelhart *et al.*, 1986).

3.5 Topological self-organization in CALM modules

3.5.1 Self-organizing maps

In this section, we will describe a simple variant of CALM that can be shown to have the capacity to arrange its representations (i.e. the stimuli represented by the R-node categories) in such a way that they reflect the topology of the stimulus input space. We have just started to investigate the many variants of CALM that have this property. Here, we will only report on some preliminary studies, and put forward a hypothesis concerning self-induced processes of local and global orderings in cortical brain maps.

It has been known at least since the pioneering work of Brodmann at the beginning of this century that the cortex can be divided into different archi-tectural regions, each with its own specific function. But it is only after the important studies of Hubel and Wiesel in the late 1950s and early 1960s (e.g. Hubel and Wiesel, 1959, 1962, 1963, 1965), that there has been a growing interest in the *topological* organization of features on the cortex. A partic-ularly interesting phenomenon is the existence of so-called *brain maps* (Knudsen *et al.*, 1987). In these maps, the spatial arrangement of representa-tions reflects functional aspects of stimuli from the outside world. In many areas of the brain this organizational principle results in good recognizable 'pictures' of some sensory modality, although more abstract 'calculation maps' can also be found whose function is more akin to a conversion table. These latter maps may, for example, convert changes in retinal position (e.g. the image of a fly on the retina) into motor signals for changing eye position (e.g. to follow that fly visually).

As an example of a simple one-dimensional map we may consider the well-known columnar organization of representations (see, among others, the studies of Hubel and Wiesel cited above). In certain columns, for example, neurons may be found that respond maximally to *line segments* with a certain *angular orientation* at a certain *position* on the retina. Such neurons are usually arranged in a well-defined order, which is in accordance with changes in a single feature. Suppose we find a column in which line segments at a cer-tain position are detected. At, say, the left side of the column, we may encoun-ter neurons that respond maximally to small lines with a '\' orientation; when moving further to the right we may then find neurons with a '—' orientation, and still further to the right with a '/' orientation (see also Essen, 1985). Thus, we have here a simple example of a one-dimensional map, its ordering reflect-ing line segments at various orientations, graphically: \ ... — ... / (the inter-mediate orientations are left out for obvious typographical reasons). It will be clear that maps need not be one dimensional, but that by assembling columns into matrices we can obtain higher-dimensional forms. It will also be clear that when we are restricted to projections on the two-dimensional cortical surface, high-dimensional maps may have a more complicated hierarchical structure, often made visible as a striped pattern (see e.g. Nelson and Bower (1990) and Appendix B2 for a more general perspective, and Ballard (1986) for an interesting hypothesis on the relation between structure and function of topographical and non-topographical maps).

Kohonen (1990) mentions many studies that have identified areas where such maps have been found. For the visual areas, for example, we find maps for colour (e.g. Zeki, 1980) and many other visual features (see e.g. Barlow (1981) for a discussion of the structure of the visual cortex from a general perspective). For the auditory areas, among others, maps for pitches have been identified (Reale and Imig, 1980; and early studies by Tunturi, 1950, 1952). A well-known map is, furthermore, the sensory map of the human skin surface (e.g. Kaas *et al.*, 1979, 1983). This map is organized topologically, but because the three-dimensional skin surface is projected on to the two-dimensional cortex, it is distorted at many points, much like an ordinary map of a significant part of the world will always be distorted (in contrast to a globe). Also, the sensory map tends to reflect the functional organization of the various parts of the body, allocating larger areas of cortex to the important sensory areas, so that, for example, the 'lip area' on the map takes up much more room than the entire 'back area', which does contain fewer sensory receptors.

From the above examples of physiological studies it can be concluded that the existence of brain maps has been known at least for some decades. Only recently, however, has there been significant progress in modelling the emergence of such architectural structures. Much of the early work in this area was performed by Kohonen (1981, 1982, 1989a, 1990). The assumptions underlying the 'Kohonen map' are very simple, and yet the method works amazingly well. We will summarize very briefly the basic rules of the algorithm:

1. A set of nodes is arranged in, say, a two-dimensional matrix.
2. Each node has a real-valued weight vector W, with random initial values.
3. A set of stimuli is available, from which at each iteration t we randomly draw a particular stimulus (vector) S.
4. We compare stimulus S with the weight vectors W of all nodes in the matrix, to identify the 'winning node'. Call this node z. (For the comparison process we could, for example, use a Euclidean metric.)
5. We adapt the weights to node z and *to all nodes in the neighbourhood of* z: call all nodes in the neighbourhood (including z itself) Z; then (only) all nodes in the neighbourhood Z are changed, according to the learning rule:

$$\Delta W = \alpha(S - W) \tag{3.2}$$

Repeating steps (3) to (5) for a large number of iterations t (sometimes up to 10 000 or more) may result in the formation of a topological map of the input space. There are two important extra conditions. In order to let the map converge to a stable ordering of representations it is usually also necessary to carry out the following:

6. Decrease the learning parameter α slowly with time t.
7. Decrease the size of the neighbourhood Z slowly with time t.

If, for example, we have a two-dimensional array (matrix) X of 10×10 nodes, where each node has two incoming connections, and if we draw stimuli randomly from a square area, say, bounded by the points (0,0) and (1,1), then after many iterations we will find that the representations of the nodes (i.e. their weight vectors) will self-organize to reflect the two-dimensional input space. At the start of the process we have a large neighbourhood (perhaps including half of the nodes in the network), which after a fixed number of iterations is slowly decreased to the point where it includes only the winning node z. Similarly, at first we have a high learning rate α, which is also slowly (e.g. linearly) decreased according to a preset low value. In the resulting two-dimensional map, we may, for example, expect node $x_{1,1}$ to have weight vector [0.05, 0.05], node $x_{1,5}$ to have [0.05, 0.45], node $x_{1,10}$ to have [0.05, 0,95] and node $x_{5,5}$ to have [0.45, 0.45]. Of course, the entire map may also be rotated 180° or 90° (to the left or right), because there is no preference for orientation. The matrix of nodes (see step 1) may also be connected at the sides (i.e. a torus structure). Many variations on this theme are possible. We could, for instance, offer the same stimuli not to a matrix, but to a one-dimensional array of 10 nodes (again with two incoming connections). In that case, the emerging representations in the one-dimensional array would attempt to cover as well as possible the two-dimensional structure of the input space (i.e. of the square). The reader is referred to Kohonen (1989a, 1990) for many examples and a thorough exposition of the possibilities and limitations of this approach.

Kohonen maps have been applied to many problems. Kohonen (1988) himself has succeeded in constructing a well-performing system for speech recognition where vectorized speech sounds are presented to a (hexagonal) Kohonen map. This high-dimensional input (15 elements) results in a 'phonotopic' map where similar speech sounds are grouped together. A spoken word, presented to such a map, leads to a trajectory over the map, activating those nodes that represent the constituent phones of the word. In a similar vein, Morasso *et al.* (1990) have developed Kohonen maps showing self-organization of an allograph lexicon, a graphotopic map where each node represents a stroke in a sequence of movements in the production of cursive script. Ritter and Kohonen (1990) describe a method whereby words, presented to a Kohonen map as parts of simple three-word sentences, self-organize on the basis of meaning. They call these semantotopic maps. Meaning relations can arise in these maps indirectly, through similarities in the contexts in which these words appear. Verbs like 'speaks', 'eats', 'likes', 'buys', 'hates', etc., are grouped together on the map, because the system has encountered these words in similar contexts, such as in '*Jim* speaks well', '*Jim* eats often', '*Jim* eats bread', or '*Mary* likes *meat*', '*Mary* buys *meat*', and 'horse hates *meat*', 'Bob buys *meat*' (where similar contexts across the verbs are indicated by italics). No physiological evidence for semantotopic maps has yet been found, although some studies suggest that higher processing stages in the brain are organized on the basis of the logical properties of their input items (e.g. Ojemann, 1983). For the moment, however, we must view the simulations on semantotopic maps by Ritter and Kohonen as a challenging hypothesis that may direct further investigations into the structure of the brain.

None of the studies cited in the previous paragraph aims at (directly) modelling the neurophysiological reality. They merely use the self-organizing map either as a convenient (learning) pattern recognizer, or for the exploration of brain processes yet to be discovered. Other studies, however, aim more directly at modelling the emergence of brain structures. Obermayer *et al.* (1990a), for example, performed a large-scale simulation (16 384 neurons, 800 tactile receptors and 13 107 200 adaptive connections) of the formation of a somatotopic map of the surface of the inner hand. The authors report that the model network adapts upon the partial deprivation of sensory input (i.e. after lesioning a finger) in the same way as is found in experiments. Another set of simulations by the same group (Obermayer *et al.*, 1990b) investigated the formation of a hierarchical topographical map with two stimulus features: (i) position in visual space and (ii) orientation. The two studies show convincingly that Kohonen maps predict many aspects of the organization of the architecture of the cortex.

The working of the Kohonen map is based on a *local selection principle (i.e. a winner-take-all mechanism) that structures a search process initiated and driven by two noise sources*: (i) initial random weights and (ii) randomly presented stimuli, respectively. This very much resembles the basic principles of the CALM module. In addition to the local selection principle, however, the Kohonen map has what we would like to call a *sublocal ordering principle: the neighbourhood*. It is the latter aspect of the system that enforces a topological structure: the sublocal interaction of representations causes a map to self-organize, so that its representations reflect a global ordering of the stimulus space. From the same basic principles (a noise-driven search process structured by local and sublocal interaction) Linsker (1986a–c, 1988) also developed a theory of self-organization. He calls his theory a *modular self-adaptive network* (Linsker, 1986a), because he considers the layers in which nodes self-organize and become ordered according to certain functional characteristics as modules. These modules may be arranged in streams and hierarchies. A neighbourhood in his theory is defined as a limited projection of a certain layer (or module) A to another layer B, i.e. only a small area of layer A gives input to a node in layer B, whereby the density of the connections follows a Gaussian-like distribution (many connections in the centre, few at the outskirts of the area). Although his work comprises many different aspects of self-organization and is not limited to topological ordering, one of the phenomena studied is the spontaneous formation of oriented, spatial-opponent nodes responding to, for example, on–centre–off–surround stimuli (Linsker, 1986a). These nodes may, under certain additional local constraints, further organize into 'orientation columns' (Linsker, 1986b–c) such as identified by Hubel and Wiesel (1962; see also the example discussed above).

We may thus conclude that the same basic principles reoccur in quite different models, and that they appear to be central to the concept of self-organization. Extending a local selection principle, such as winner-take-all competition with a sublocal ordering principle, results in a topological ordering of representations. In the remainder of this section we will show that by introducing a sublocal ordering in the CALM module it also performs topological ordering of input data.

3.5.2 Topological self-organization in CALM

A sublocal ordering principle can be introduced in CALM in several ways. In Chapter 7, we will discuss variants of CALM that are more biologically plausible (i.e. with fewer V-nodes than R-nodes). As will be shown, these variant modules can easily exhibit topological self-organization. Here, we will concentrate on the effect of the introduction of a neighbourhood through *graded recurrent lateral inhibition* in the cross weights (see Figure 2.3). In CALM, these weights have a constant low value (e.g. -10.0), which underlies the veto characteristic of the V-nodes. We can arrange the R–V-node pairs in a CALM module in a one-dimensional array and define the neighbourhood of an R–V pair by the diminished inhibition of 'near' R–V pairs, for example a linear increase in inhibition of -1.25 for every next R–V pair (starting with -1.25 for the down weight, see Figure 2.3). This is shown for CALM (constant cross weights) and for CALSOM (graded cross weights) in Figure 3.5. From here on we will refer to CALM modules with graded cross weights as CALSOM modules (from Categorizing And Learning Self-Organizing Modules, cf. Kohonen's Self-Organizing Maps). In the CALSOM modules discussed here

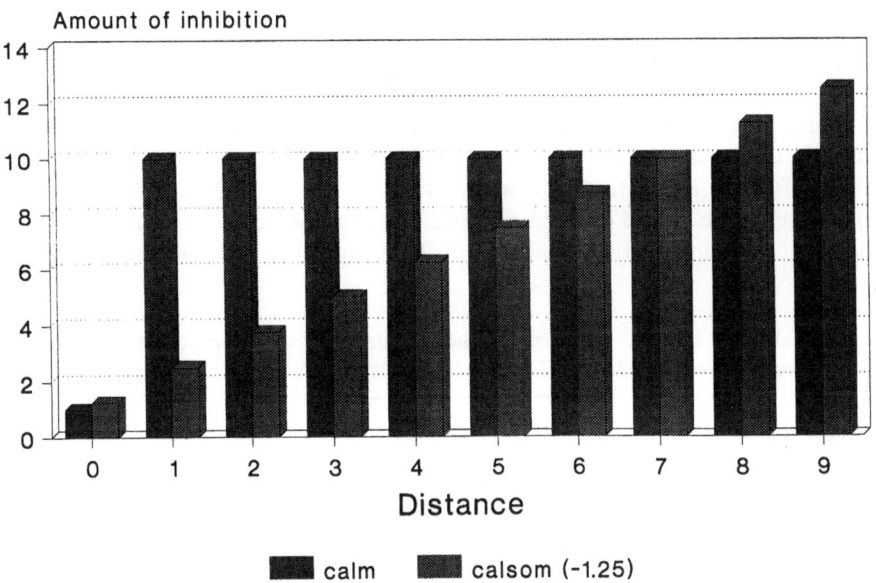

Figure 3.5 Graded (recurrent) lateral inhibition in CALSOM. The cross weights' inhibition decreases by an amount of -1.25 with distance (measured in R-nodes from the current R-node). The constant value of -10.0, used for the cross weights in CALM in this work, is shown for reference. Note that the down weight is also changed in CALSOM (to -1.25, in the figure at distance 0).

only one-dimensional topologies are considered, without wraparound (i.e. a line, rather than a ring).

It may not be obvious to the reader why the introduction of a neighbourhood with diminished inhibition causes similar pattern representations to be grouped together. Suppose, as an example, that a pattern S has been categorized at node 5 in some CALSOM module. If we now present a pattern T to this module that is *very* similar to S, it will of course be categorized at node 5. But if it is only 'moderately similar', a competition process will be initiated. We know that, during the competition process, node 5 will be activated somewhat (because T moderately fits representation 5). V-node 5 will thus also be activated, and through the graded cross weights the R-nodes close to R-node 5 will receive *less* inhibition than R-nodes at some distance from node 5. So we could expect that, for example, pattern T is categorized at node 6 or 7, rather than, say, node 10, 11 or higher. This is the basic dynamic mechanism that groups similar patterns closely together.

In order to verify whether the introduction of sublocal neighbourhoods does indeed induce an ordering of input patterns in CALM, we ran a small series of simulations. A set of 10 patterns was constructed, each consisting of 12 values. The patterns, shown in Table 3.3, have a natural ordering of 1 to 10. These patterns were presented to a CALSOM module of size 10. Graded recurrent lateral inhibition was as in Figure 3.5. A hundred patterns were randomly drawn from the set (i.e. not all patterns were presented the same number of times). Each pattern presentation lasted for 30 iterations. In between presentations, all activations (but not the weights!) were reset to zero. After these hundred presentations all patterns 1 to 10 were presented once more in order to identify which pattern was represented by which R-node. The results are shown for five replications in Table 3.4. From this table, it can be seen that the patterns are indeed ordered on a one-dimensional scale. In all replications, however, we find certain R-nodes representing more than a single pattern.

Table 3.3 Pattern set used in the simulations with topological self-organization in CALM (CALSOM modules).

No.	Pattern											
1	1	1	1	0	0	0	0	0	0	0	0	0
2	0	1	1	1	0	0	0	0	0	0	0	0
3	0	0	1	1	1	0	0	0	0	0	0	0
4	0	0	0	1	1	1	0	0	0	0	0	0
5	0	0	0	0	1	1	1	0	0	0	0	0
6	0	0	0	0	0	1	1	1	0	0	0	0
7	0	0	0	0	0	0	1	1	1	0	0	0
8	0	0	0	0	0	0	0	1	1	1	0	0
9	0	0	0	0	0	0	0	0	1	1	1	0
10	0	0	0	0	0	0	0	0	0	1	1	1

Table 3.4 Simulation of five independent replications with CALSOM. The pattern numbers (rows) coincide with those in Table 3.3. For five replications (columns) the R-nodes have been listed that were found to represent the pattern. R-nodes have been numbered 0 to 9. Note that a single R-node may represent more than one pattern. Locally deviating suborderings have been indicated in italics.

Pattern	Replication				
no.	1	2	3	4	5
1	8	7	5	*9*	*0*
2	8	6	4	0	*0*
3	7	5	4	0	9
4	7	4	3	2	8
5	6	3	2	2	8
6	6	3	1	3	7
7	5	0	9	4	6
8	4	*8*	*8*	5	4
9	2	*8*	7	6	2
10	1	*9*	7	7	2

Discrimination is thus not yet optimal. This may be caused by the fact that local competition is diminished, which increases the chance that representations are merged.

Local deviations (suborderings) from global orderings are found in replications 2 to 5. (In replications 4 and 5, they consist of single, isolated representations.) Kohonen (1990) reports a similar lack of global ordering, if the neighbourhood and learning parameter are not adjusted according to the right schedule (i.e. steps 6 and 7 above). Of course, in the simulations with CALSOM no such global schedule was used. The module is left entirely 'on its own'. The objective of these simulations is merely to show the possibility of inducing topological order in a CALM module. The results are clearly supporting the feasibility of this approach, but more thorough research is necessary to establish the precise relations between such parameters as inhibition gradient and local topology (i.e. the *form* of the neighbourhood), the nature of the input stimulus space (e.g. high or low dimensional, high or low similarity of inputs), and the resulting ordering. We might also investigate the effect of stacked CALSOM modules. Such models may be compared with Linsker's modular self-adaptive network (see above).

Our research on CALSOM is still in progress. At this moment we can report on some further, preliminary simulations that study one possible reason for the problem of lack of global ordering. As noted, Kohonen (1990) attributes part of these problems to an improper neighbourhood 'shrinking schedule'. If, in a Kohonen map, one starts out with a neighbourhood that is too small, no global initial ordering of representations will take place. In CALSOM the neighbourhood is defined by the inhibition gradient along the R−V pairs (i.e. diminishing cross weights, see Figure 3.5). Although this gradient is fixed, it will cause a *bubble of competition* that will decrease in size with subsequent pattern presentations. At first, even distant neighbours of a given R-node will compete for a representation, because initially many R-nodes have similar, and unordered, representations. The bubble of competition will at this time be fairly large. After sufficient learning has taken place, however, distant neighbours will have been tuned to patterns further removed in pattern space, and they will, therefore, no longer take part in the competition process. Only close neighbours will compete. Learning, in CALSOM, thus has the effect of decreasing the size of the bubble of competition, even though it is induced by a fixed inhibition gradient (neighbourhood).

We have run simulations to verify this account of the ordering process in CALSOM. The same input patterns from Table 3.3 were used. A hundred patterns, randomly drawn from the set, were presented for 30 iterations each to a CALSOM module of size 10. Activations were initialized in between presentations. The details of these simulations coincide with those of the simulation described above. We were mainly interested in how the bubble of competition caused by a given pattern S_i would decrease with subsequent presentations of that pattern S_i. At each pattern presentation of S_i we therefore recorded the cumulative activation of each R-node in the module. This was done by simply calculating the sum total of the activations of a given R-node j over all 30 iterations. Let us call the sum total for each R-node j, T_j . For R-nodes that did not take part in the competition, of course, T_j was very low. For the winning R-node T_j was highest.

In this simulation, the initial topological ordering, at the first 25 random presentations, was (8 8 7 6 *4 5* 2 1 0 0) for patterns 1 to 10, respectively. Later on, patterns 5 and 6 swapped their representations, while the representations of R-nodes *0* to *2* shifted to the right, the previously uncommitted R-node *3* being assigned to pattern 7. These two local reorderings resulted in the strictly descending global ordering (8 8 7 6 5 4 3 2 1 0), again for patterns 1 to 10, respectively (see Hartsuiker (1991) for additional details of the local reordering process). Except for the double representation of R-node *8*, which represents both patterns 1 and 2, the ordering is perfect.

In Figures 3.6 and 3.7 the bubbles of competition are shown for patterns 5 and 6, respectively (see Table 3.3). Comparing Figures 3.6 and 3.7 we may notice two effects. (i) At the second and third presentations the competition bubbles swap places in both figures. This is the local reordering of representations, described above. (ii) Especially in Figure 3.6, we see that the size of the bubble of competition strongly decreases with subsequent presentations, at first ranging from R-nodes *1* and *2* to R-nodes *7* and *8*, and at the eighth presentation being limited to R-nodes *3* to *6*.

From this preliminary simulation we may thus conclude that in CALSOM we find some evidence that a locally and autonomously induced shrinking of the bubble of competition takes place. Similarly, we are presently investigating

cumulative activation

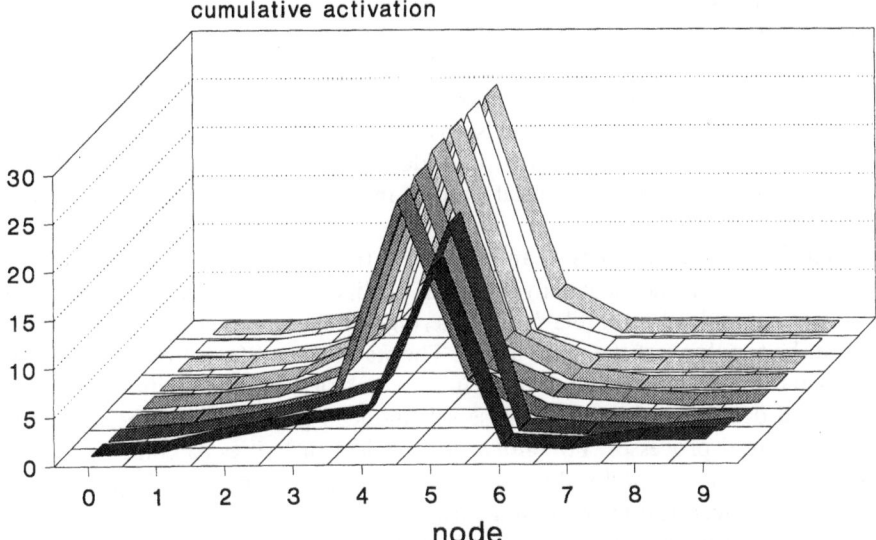

Figure 3.6 Bubbles of competition caused by pattern 5 (see Table 3.3) at eight subsequent presentations. The first presentation is shown at the front (dark band) and the eighth presentation is shown at the back (light band). The presentations were randomly preceded by zero or more other pattern presentations. The height of the graphs represents the sum total of activations of each R-node, measured over the 30 iterations of a single pattern presentation (i.e. the statistic T_j, see text).

cumulative activation

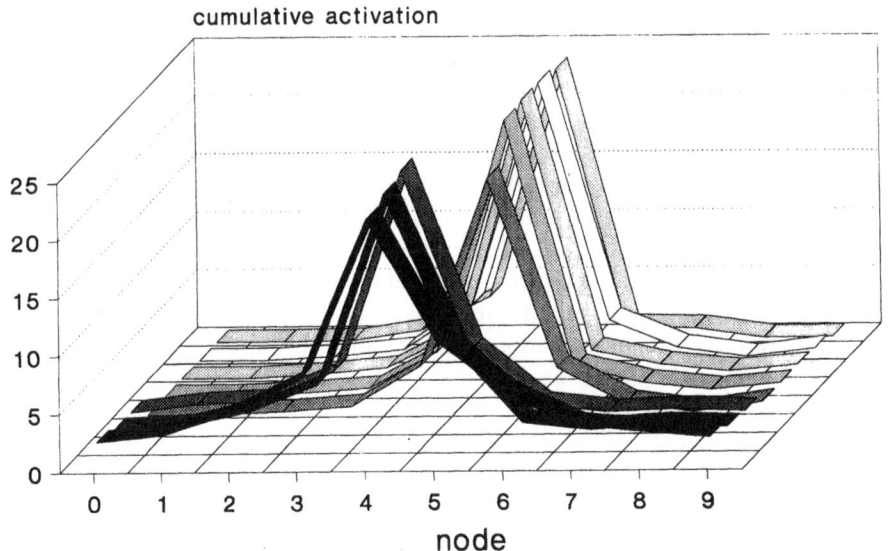

Figure 3.7 Bubbles of competition caused by pattern 6 (see Table 3.3) at eight subsequent presentations. See Figure 3.6 for an explanation.

whether a CALSOM module has a self-induced decreasing learning parameter, which was found for CALM. At this moment, we cannot report any firm conclusions on this. From the results of the work on CALSOM so far, we can conclude that it is able to perform a local ordering of input data, but that for a *global* ordering, additional provisions of a superlocal nature may be necessary. This may be in the form of a global neuromodulator that could for instance affect either the arousal process in a global way or the cross-weight gradient.

The following hypothesis can be derived from the results of the CALSOM simulations for the neurophysiological reality: lateral interactions between neurons may lead to stable *local* orderings of input patterns, in accordance with the topology of the input space. The stability and extent of the orderings is determined by self-induced shrinking bubbles of competition and self-induced decreasing learning. These self-induced processes may be related to local arousal processes. For *global* topological orderings to come into existence, however, a superlocal process may be necessary, possibly in the form of minimal concentration of some neuromodulator.

A recent study by Juliano *et al.* (1991) reinforces the plausibility of the latter hypothesis. One of their research questions addressed the role of cortical acetylcholine (ACh) in the emergence of maps in the somatosensory cortex (identified through metabolic activity), and the extent to which ACh participates in neocortical map rearrangement following amputation. Their results indicate that *cholinergic innervation is a requirement for the rearrangement of cortical maps*. Cortical maps of the forepaw of cats with amputated second digits were found to become significantly rearranged. They investigated the pattern of metabolic activity in the area previously representing the amputated digit. When stimulating the adjacent digit the pattern of metabolic activity in the somatosensory cortex was found to be dramatically expanded when compared with the opposite (normal or non-amputated) hemisphere. As Juliano *et al.* (1991, p. 780) report: 'In contrast, experiments in which the somatosensory cortex was depleted of acetylcholine and the animal received a similar amputation led not to patterns of expanded metabolic activity, but rather to reductions in the evoked metabolic activity'.

The authors also link ACh depletion to a reduction in *stimulus-evoked activity*. As was already shown by Hubel and Wiesel (1965) perceptual experience is often an important driving factor in emerging neurophysiological representations. Self-organization in the brain will thus generally be stimulus evoked, even though the stimuli may be generated internally (as Linsker (1986a–c) has shown). We could thus envision a process whereby high ACh levels at first induce stimulus-evoked (or self-induced) global orderings, and that by gradual depletion in the system, only stimulus-evoked local orderings could be made (e.g. for fine-tuning of representations). The gradual depletion of ACh may be an autonomously induced process working at a larger timescale in the developing animal than the self-induced arousal processes modelled in CALM and CALSOM. Full depletion would lead to a total absence of neural plasticity, and we would thus rather expect the system to stabilize at a certain low level of ACh. Of course, this hypothesis is speculative, and it may not do full justice to the wealth of data available on the processes of neural plasticity. It is our aim to extend the simulations with CALSOM-type models and to investigate further the mutual influences of the interlocking processes of globally and locally induced processes of self-organization.

PART II
Application

PART II

Application

Chapter 4

Psychological Models

In Part II, CALM is used as a basic building block for larger neural network models. In this chapter two psychological models will be discussed. In the following chapter some aspects of practical applications of CALM are studied. The first model, described in Section 4.1, shows how the context effect for letter recognition may be learned in an unsupervised manner. The second model is used in a series of simulations of experiments in memory psychology. It is able to account for results obtained from various experiments on implicit and explicit memory, both with amnesics and with normal subjects. The model and simulations are discussed in Section 4.2. Several other projects that apply CALM to psychological modelling are in progress. These projects are concerned with the development of early speech perception and production, with categorization and generalization, with implicit and explicit measures of emotional behaviour, and with additional aspects of memory (including list learning in models that possess an articulatory loop). We will report on this work elsewhere.

4.1 Learning the word-superiority effect for letter recognition

An important property of CALM networks is their ability to organize input–output relations without specifying the desired output. Such learning is generally called unsupervised or regularity learning (e.g. Rumelhart and Zipser, 1985; Schneider, 1987). CALM provides a building block for building larger networks that perform such learning by successive categorization in modules of subsequent layers. It may thus be seen as a hierarchical clustering device discovering statistical regularities in the input patterns. Such a pattern hierarchy, for example, appears to lie at the basis of the well-known letter-recognition model of McClelland and Rumelhart (1981). The connections from the first layer to the second layer transform line features on particular positions to letters on these positions. The subsequent layer of connections leads to the composition of four-letter words. The network consists of nine modules: four for all positions at the first and second layer and one for the

59

words at the third layer (see also Figure 1.1). The model shows an advantage in the recognition for masked letters presented in word contexts above masked letters in irrelevant or non-word contexts.

Critical features of the model are the hierarchical organization of letters and words and bidirectional connectivity between word and letter level. These features especially have recently received much criticism (e.g. Massaro, 1988; Mewhort and Johns, 1988). Our present simulation is, however, not concerned with the psychological plausibility of the model *per se*, but with the question of how such a model can come into being; a question that is not answered by McClelland and Rumelhart (1981). One of the reasons for choosing the McClelland and Rumelhart (1981) model to illustrate the development of representations in a multi-modular network is that Grossberg (1987b) has criticized just this aspect of the model. According to his critique, the model is unable to self-organize, because it would be unstable in a learning mode. We show here that a network consisting of five CALM modules and four input modules may self-organize to form stable letter and word representations, while preserving the two critical features of the McClelland and Rumelhart model. A major difference between the present model and the McClelland and Rumelhart model is the absence of interlevel inhibitory connections. These connections have been held responsible by Grossberg (1984) for the inability to self-organize and may, therefore, explain the deviation from his prediction.

The implementation of the letter-recognition model was restricted in the number of words and letters to be represented. Dutch words of 24 letters were selected. These words had a considerable number of letters in common on the same positions. In fact, 10 word pairs had an overlap of three letters (i.e. 75% overlap, or more). Due to this overlap, only a limited number of different letters were required on each position. On the first position seven different letters could be presented (A, C, D, E, H, N and T), on the second, third and fourth only five (A, C, E, H, T; A, E, H, T, N; and A, D, E, T, N, respectively). Letters were presented to the network as quasi-features by activating nodes in a 5×7 grid with an activation of 0.5. The nodes were selected by drawing the letters in the grid. The number of activated nodes in the grid varied from eight to 16.

The full network consists of nine modules: four input modules (of 35 nodes) at the first level, four CALM modules (of 7, 5, 5 and 5 nodes, respectively) at the letter-position level, and one CALM module (of 20 nodes) at the word level. Every input module is fully and unidirectionally connected to only one CALM module at the next level representing a letter on a particular position. The four letter-position modules are, in turn, fully and bidirectionally connected to the single CALM module at the third (word) level. In comparison with previous simulations two minor deviations can be found in this simulation. First, a larger value for the decay parameter (0.25 instead of 0.05) has been used. This is not an essential modification, since the previous simulations will also work with this value for the parameter, though convergence times will be altered somewhat. The advantage of larger values is that they help further to prevent the occurrence of reverberatory loops between R-nodes, when there are bidirectional connections between two CALM modules. The second deviation is the inclusion of an additional Veto-node in the CALM module at the word level. This node is activated at the onset of presentation of the word and will inhibit all nodes in the module until its activation has decayed. It receives no activation from other nodes. The V-node preactivation gives the convergences in the letter-position modules a headstart over the convergence in the word module, so that in the learning phase an early 'word-superiority' effect will not disrupt the formation of appropriate representations for new words when other words are already learned. The preactivation has been chosen low (0.4), so that for all practical purposes it will have ceased to influence convergence in the word module after about 15 iterations, when it has decayed to about 1% of its initial value.

Presentation of the words took place in two phases. First, all letters were made known to the model by presenting them separately to the model. Twenty-two letters (all letters at all possible positions) were presented 20 times for 50 iterations each. Except for the special V-node preactivation no other activations (e.g. of certain R-nodes) were presented to the network. Learning was, thus, unsupervised. This resulted in stable and separate representations for the letters on the positions. Following the learning of letters, the 20 words were presented 60 times for 100 iterations each, also in the absence of supervision. Though sufficient R-nodes were available in the word module not all words were separated. Two of the 10 word pairs that had three letters in common were not separated. For instance, the Dutch words 'acht' and 'echt' were represented on the same R-node. Letters at the first position also showed considerable overlap (e.g. 'a' and 'e' in the example), which made discrimination even harder. In Section 3.3 it was shown that for the separation of two non-orthogonal patterns by a two-node CALM module a minimum Euclidean distance of 0.7 was required. This minimum distance for the more general case is, however, a complicated function of the size of the pattern and the size of the module. When the input pattern consists of four activations and the module has 20 R-nodes, a difference in only one of the activations may not always be sufficient to separate the patterns. It is important to stress that otherwise the categorization was completely stable. Presentation of either a letter or a word always led to activation of the appropriate R-nodes. This stability proved to be so robust that it was in fact difficult to find conditions in which the model would make errors, as is required for finding a word-superiority effect. In order to score the test results all representations for both letters and words were recorded.

Although we distinguish between a learning and a test phase, learning was not disabled in the test phase. The distinction can only be made on the basis of the stimuli presented to the network, not on the basis of some internal condition of the network. To test for the word-superiority effect we presented poorly recognizable letters either in a word context or in the context of an anagram consisting of the same letters. Such a design minimizes confounding from other factors, which may enhance the difference between word and non-word contexts (D.J.K. Mewhort, personal communication). Moreover, under these conditions a small but consistent word-superiority effect has been found with human subjects.

Finding conditions that resulted in the poor recognition of single letters proved very hard, because even severely degraded letters could be identified correctly by the model. Eventually, we reduced input activations of the degraded letters from 0.5 to 0.0005 to produce a significant amount of errors. Eight target words were selected randomly of which one letter would be degraded. For every position two different letters were degraded. Eight anagrams with the same degraded letters in the same position were also prepared. Testing of the targets and the anagrams was replicated 25 times.

Degraded letters were recognized better in the word context (proportion correct 0.79) than in the anagram context (0.54). Recognition time was also reduced in the word context (23.0 iterations on average) compared with the anagram context (25.4 iterations). There was no difference in reaction time for the incorrectly recognized letters (29.6 iterations for both word and anagram context). It can be concluded that with some slight modifications a learning version of the McClelland and Rumelhart (1981) letter-recognition model can be implemented in terms of CALM modules, while retaining the critical features of the model.

4.2 Modelling human memory*

4.2.1 Neural networks and memory models

Despite its excellent suitability for connectionist modelling (see e.g. Kihlstrom, 1987) no network model simulating actual implicit memory experiments has appeared so far. With the model presented here, two extensive simulations of dissociation effects between explicit and implicit memory tests have been carried out. The simulations show differential effects of word frequency and amnesia in implicit and explicit memory tasks which are qualitatively similar to empirical findings. It should be stressed from the outset that the network model used is a relatively simple one. Therefore, we did not endeavour to attain maximal quantitative precision. Our aim is restricted to showing that the model, which is based on a single set of assumptions and parameters, is capable of capturing some of the intriguing dissociation effects found in explicit and implicit memory tests.

* The work reported in this section is part of a project sponsored by NWO (grant no. 560-259-027 to G. Wolters). The primary researcher on this project is R. Hans Phaf. A more extensive and more detailed account of these simulations can be found in Phaf *et al.* (1990a) and in Phaf (1991). The latter work also includes full reports of a series of experiments in implicit and explicit memory.

In the first section, a theoretical explanation of the differential effects found in implicit and explicit memory tasks is discussed. In later sections, some general ideas are presented about the structure and functioning of a network capable of simulating such dissociation results, and a detailed account of the network model will be given, together with procedural details of the simulations. Finally, in Section 4.2.6, the network model used here will be evaluated and some extensions and further results will be discussed.

4.2.2 Implicit and explicit memory

Graf and Schacter (1985, p. 501) define implicit memory as follows: 'Implicit memory is revealed when performance on a task is facilitated in the absence of conscious recollection'. Thus, implicit memory is not the result of an active retrieval process on the part of the subject. An implicit memory task rather evokes an increased probability of a response, or a decrease in latency, following prior presentation, even if there is no awareness that the prior presentation is being tested. Implicit memory tasks can be contrasted to the traditional memory tasks, such as free recall, cued recall, and recognition, that do require an active recollection of the learning episode. In these tasks an explicit reference to the study or experimental context is always required in the instruction given to the subject. Without such a reference no recollection can take place, because it is undetermined what should be remembered. Therefore, tasks like, for instance, free recall or cued recall are called explicit memory tasks. Most computational models for explicit memory performance (e.g. Raaijmakers and Shiffrin, 1980, 1981) assume that the subject uses such a context cue for recalling the items learned during the experiment. For an implicit task, a context cue does not seem to be required, because a specific learning episode need not be referred to. Typical implicit tasks are threshold identification, lexical decision, category exemplar generation, and word completion.

A good example of an implicit memory task is word completion. In a word-completion task the subject is provided with word stems, each consisting of the first few letters of words, which may or may not have been presented before. For instance, if one of the presented words was WINDOW, the word stem would be WIN. In the completion task, the subject is instructed to complete the stem with the first word that comes to mind. The advantage of completing presented words, like WINDOW, over non-presented words, like WINNER, forms a measure for implicit memory. Generally, experimental manipulations prevent the subject from becoming aware of the memory test character of this task. The advantage of old words over new words, observed in these tasks, has also been referred to as 'repetition priming' (Cofer, 1967; Murrel and Morton, 1974).

Theoretically, the distinction between implicit and explicit memory is important since experiments using implicit and explicit tasks as measures of memory for earlier presented material reveal a clear dissociation between the two forms of memory. For instance, Jacoby and Dallas (1981) and Graf and Mandler (1984) have found that the level of processing of the presented material affects explicit memory performance, but does not affect implicit memory performance. Schacter (1987; see also Richardson-Klavehn and Bjork, 1988; Roediger, 1990) reviews a large number of experimental findings that support such a dissociation. Furthermore, a striking argument for this dissociation can be found in patients with anterograde amnesia who are impaired on explicit tasks but show a near-normal performance on implicit tasks (Warrington and Weiskrantz, 1968, 1970). Generally, the results contradict the view that performance in implicit and explicit tasks is based on a single memory phenomenon. In particular, the notion that implicit tasks would have a different threshold for retrieval than explicit tasks is falsified by these results (Schacter, 1987).

There are two main theoretical explanations offered for the dissociation between implicit and explicit memory tasks. The first explanation assumes multiple memory systems that are separate and functionally different. The second postulates two different processes within a single memory system. An influential example of a multiple memory explanation of implicit and explicit memory is the declarative–procedural memory distinction made by Squire and Cohen (1984; see also Winograd, 1975). In procedural memory information is stored about skills and procedures, whereas in declarative memory information about specific facts or episodes is stored. Declarative memory is explicit and accessible to conscious awareness. Procedural memory, however, is implicit and accessible only through performance.

Another multiple memory view is the episodic–semantic memory distinction proposed by Tulving (1972), which may also be used to explain the dissociation. Episodic memory deals with unique, concrete, temporally dated events, while semantic memory involves general, abstract, timeless knowledge that a person shares with others. Explicit memory would then reveal properties of episodic memory and implicit tasks more of semantic memory. Semantic knowledge has to be learned, however, and the only way this learning can take place is through episodic experience (see also Stanfill and Waltz, 1986). So, it can be argued that general semantic knowledge may gradually be abstracted from more specific episodic knowledge (Wolters, 1984). Of course, once semantic knowledge has been established, it will be used in interpreting and constructing further episodic experiences, so that personal episodic memory also may be based on more general semantic knowledge of the world (Tulving, 1983).

Multiple memory theories are incomplete without a specification of how the distinct memory systems interact. Such theories may be able to explain dissociation results by referring to the separate memory systems, but they fail to describe how information from these systems may cooperate to produce coherent behaviour. Through precise specification of such an interaction, however, the sharp distinction between the multiple memories may vanish. For example, as was argued by Wolters (1984), it is possible to see episodic and semantic memory as two functionally different aspects of a unitary storage system.

From a methodological point of view, a multi-process theory in a single memory system is more parsimonious and may be preferable (McClelland and Rumelhart, 1986b). The connectionist model proposed here gives a multi-process storage account of the dissociation. According to Richardson-Klavehn and Bjork (1988) such an account of implicit and explicit memory may have the great advantage of integrating, ultimately, within one theoretical framework the storage and retrieval of personal experiences and the organization and retrieval of more general (conceptual, factual, lexical, perceptual or procedural) knowledge.

A multi-process hypothesis concerning the implicit–explicit memory dissociation, which lends itself well to implementation in neural networks, has been proposed by Graf and Mandler (1984; see also Mandler, 1980). According to Graf and Mandler the differential effects in implicit and explicit memory result from the existence of two different learning mechanisms: the strengthening of existing associations and the formation of new associations. In this view different memory tests are sensitive to different aspects of the underlying memory representations. An implicit memory test requires *the partial or complete reinstatement of a stimulus representation by using the relations that constitute the representation.* Performance in such a test is enhanced when a prior stimulus presentation has strengthened the internal relations by an automatic *activation* process occurring when an existing representation is accessed. Explicit memory tests require *access to a stimulus representation via retrieval routes that make use of newly formed links between the representation and information specific for the episode.* This kind of test is dependent upon the formation of new relationships in a learning process which has been called *elaboration.* The formation of new links is not seen as a consequence of mere coactivation, but is considered to result from active attentional processing: 'A specific new relationship between previously unrelated events does not appear automatically, it must receive focal attention to become encoded' (Graf and Mandler, 1984, p. 566). Attention as an intervening variable in elaboration learning also seems necessary for accommodating the evidence that items that have not been attentionally encoded can be remembered implicitly but not explicitly (Eich, 1984). A possible translation of the two processes in (abstracted) neural network terms is illustrated in Figure 4.1. Circles depict representations or part of representations and connections represent the relations between them. A full circle corresponds to an active representation. The thickness of lines corresponds to the strength of associations.

Activation is assumed to be a mechanism that only strengthens pre-existing connections. When a stimulus is presented to a network, the corresponding representation is directly and automatically activated. This activation can have a lasting effect, because the associations, embodied in the connection weights, may change as a result of the activations. Such an effect will reveal itself in the efficiency of activation of the representation on a later occasion, which may correspond to the facilitation found in implicit memory tasks.

The *elaboration process* involves the formation of new connections and

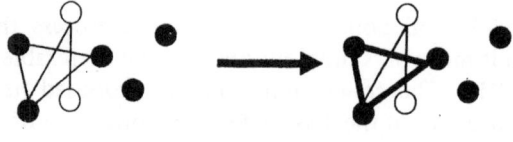

ACTIVATION

passive strengthening of pre-existing associations

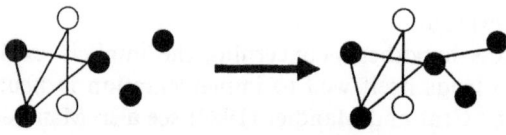

ELABORATION

active formation of new associations

Figure 4.1 Implementation of the activation/elaboration hypothesis of Graf and Mandler (1984) in neural network terms. See text for details.

results from a more active exploration controlled by focal attention. Attentional processes may very well be implemented in neural networks by selectively enhancing the activations of some representations (e.g. Phaf *et al.*, 1990b). There is, however, the problem of which representations have to be selected for such enhancement. This selection should, of course, not be made beforehand, but has to be performed by the model. In the present network model only aspecific attentional activations are provided to a module, with the actual selection of the specific representations being executed by the model itself. In the human information processing system, attention can be directed by many causes. For example, saliency of stimuli, instructions, or intrinsic motivation of the subject can all direct attention. In the present network model only the relative novelty of a stimulus (or combination of stimuli) is taken as the factor evoking 'attention' to the processing in a particular module. Increased attention results in the transition of activation learning to elaboration learning.

The network implementation of activation–elaboration is actually somewhat more general than the formulation by Graf and Mandler (1984). First, the dichotomy suggested by the psychological description of the two learning mechanisms is instantiated in the model as a continuum. Activation and elaboration learning are the poles of this continuum, ranging from the mere strengthening of existing representational connections to the formation of new representational connection patterns. Secondly, elaboration learning has not been implemented as the formation of completely new connections, but as the formation of a *differentiated set* of connection weights in bundles of connections that formerly had equal, non-zero, connection weights. In winner-take-all networks such aspecific connection schemes correspond to the absence of a winner (i.e. an already established representation), so that only with the emergence of differentiated weights may connections between representations develop (resulting in the formation of new representations). Moreover, as a result of the continuum between both learning processes in the network, elaboration may also lead to a facilitation of implicit memory. On the one hand new associations formed by elaborative learning may show up in implicit tasks, such as the priming of new associations (see Graf and Schacter, 1985), and on the other hand elaboration may result in some additional strengthening of old representations (for a detailed discussion of these extensions, see Phaf *et al.*, submitted).

4.2.3 ELAN-1: a network for implicit and explicit memory

One of the central issues in this book is the assertion that the modular construction of models offers several advantages above non-modular networks. In modelling memory, modularity may be especially appropriate, because it may contribute further to the dissociation between implicit and explicit memory. For example, it may give a more specific account of the 'representational components' proposed by Graf and Schacter (1987). Different modules may be addressed by the implicit and explicit retrieval information. For instance, the presence of an effect of modality (i.e. the sensory modality relation between storage and test) in implicit tests but not in explicit tests (Schacter and Graf, 1989; Phaf *et al.*, submitted) strongly suggests that such a factor affects the dissociation.

We have called the family of network models constructed from CALM modules for performing implicit and explicit memory tasks ELAN (ELaboration and Activation Networks). The simplest model will be discussed here and is called ELAN-1. Other members of this family (ELAN-2 and ELAN-10) will be discussed elsewhere.

The design of the model has been inspired by existing models for long-term memory, like, for instance, SAM (Search of Associative Memory; Raaijmakers and Shiffrin, 1980, 1981). Although this (computational) model simulates explicit memory very well, it has not been applied to implicit memory tasks and presumably would not be able to produce a dissociation between implicit and explicit tasks. SAM, however, has some properties that are very well suited for our purposes:

1. It distinguishes between contextual information and item information. Contextual information concerns the temporal–contextual setting where the item information has been learned (e.g. Raaijmakers, 1987). In SAM, item information stored in memory is retrieved with the help of cues: pieces of information that have been associated with the information to be retrieved (e.g. a context cue for retrieving items that have been stored under that context). In similar vein, retrieval in the ELAN-1 model is driven by

externally presented cues. Task performance in an experimental setting is, of course, mainly controlled by the instructions. Although the instructions provided with a task are not incorporated in the present model, they can be added to it in a consistent manner (see e.g. Phaf *et al.* (1990b) for a discussion of how instructions can be presented to a network model).

2. The sampling and recovery procedures used by Raaijmakers and Shiffrin (1980, 1981) have been translated into the network architecture and allow for a similar kind of probabilistic and sequential response production. This will be explained in more detail later on.

3. Learning during retrieval, which is also assumed by Raaijmakers and Shiffrin, is an automatic consequence of the unsupervised CALM learning procedure.

A word-completion task was chosen as the implicit task and a free-recall task as the explicit task. In order to be able to specify input and output representations in the ELAN-1 model, a number of simplifying assumptions were made. To simulate free recall and word completion, three kinds of input representations were made available to the model: *word beginnings, word endings* and *environmental contexts*. (Since the model is not intended to model low-level word recognition, there is no provision made for encoding the ordering of word parts.) The only output representations are *complete words*.

Figure 4.2 shows the ELAN-1 network used for the simulations of implicit and explicit memory phenomena. The core of the network is formed by two

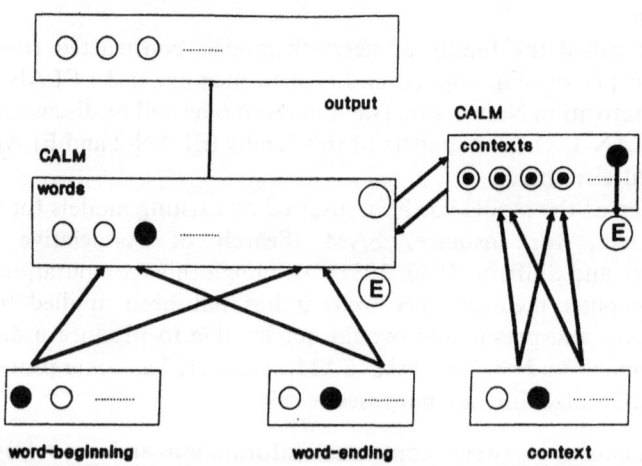

Figure 4.2 The full ELAN-1 model with schematic wiring pattern. Between CALM and input modules full unidirectional connectivity exists between input and R-nodes. Between CALM modules there is full bidirectional connectivity for the R-nodes. The output module is connected by one-to-one connections from the R-nodes of the CALM module to the output nodes. Full or partly full nodes show possible activation states.

CALM modules (labelled 'words' and 'contexts'). Five additional modules are used to represent input to and output from the network. In the word– and context–CALM modules, every node stands for the representation of a potential pattern presented to those modules from the input modules. The actual CALM modules used consist of 25 word Representation-nodes and seven context Representation-nodes. The two CALM modules are linked by bidirectional (but not symmetrical) connections between all R-nodes of both modules. (This leads to 350 variable interweights between the CALM modules.) Word parts and contexts are presented to the network through the three input modules in the lower row. An input module simply holds an input pattern in a row of 'clamped' input nodes (i.e. nodes with unchanging activations). In the left input module (consisting of 10 input nodes) the activation of a node represents the presence of a word beginning, and in the middle input module (also consisting of 10 input nodes) each node represents a word ending. The activation of a pair of nodes in these two modules signals the presentation of a complete word. During learning input nodes may become associated with a single node in the word module (representing the full word). The third input module includes seven arbitrary contexts. The activation of one node in this module represents a particular environmental context of the words. Input modules are fully connected to their corresponding CALM modules by unidirectional learning connections. (This results in a further 540 learning connections.) Before learning, all variable connections have the same initial value which was taken half-way between the maximum and minimum values of these weights.

Learning takes place in the network through the modification of the weights of the unidirectional connections between input modules and word and context modules, and of the bidirectional connections between word and context modules. *Retrieval* of a word can take place in this model along two different routes, by presenting either a context or a word part as a cue. The output is, however, the same for both routes. In the present model complete words are produced as output. A special output module has been designed for producing sequential responses with the model. The output module is coupled to the word–CALM module (see Figures 4.2 and 4.3). An active output node in the output module represents the response production of a particular word. The output nodes are directly activated by the nodes in the word–CALM modules through one-to-one connections from the representation nodes to the output nodes. One V-node, which strongly inhibits all output nodes, can be found in the output module. This V-node is activated by the A-node in the word–CALM module. It blocks all response production until the competition among the word nodes has been settled. So, an output node can only be activated after a single R-node if the word–CALM module has won the competition, and the A- and V-node activations in this module have decayed sufficiently. The R-node in the word–CALM module which receives the highest excitation, of course, has the greatest *chance* of winning the competition. The output module enables the selection of a single node with a

OUTPUT-MODULE

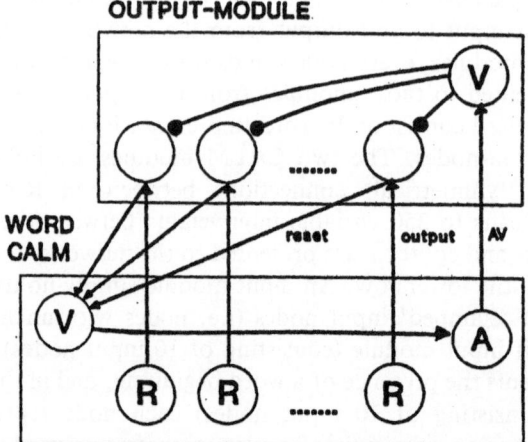

Figure 4.3 An output module consists of output nodes that receive their input through one-to-one connections from the R-nodes in the coupled CALM. The V-node strongly inhibits all other nodes in the output module. A fixed AV weight connects the A-node in the CALM with the V-node in the output module. An activation in the output module is fed back to the CALM through reset weights.

probability that is a function of the activity of the corresponding word–CALM module R-node. The combined function of a CALM module and an output module allows for a probabilistic response production similar to the sampling and recovery procedures in the SAM model (Raaijmakers and Shiffrin, 1980, 1981). It is interesting to note that these rather complex procedures can be obtained with only a small extension of the CALM module.

Connections from the output nodes to an additional Veto-node in the word–CALM module have been included to enable the production of *sequences* of different responses. In the present model, this particular V-node has also been connected to a similar node in the context–CALM module (a VV-connection, not shown in Figure 4.2). The activation of an output node thus results in a strong inhibition of all R-nodes in both CALM modules. Another connection from the V-node in the output module to the A-node in the word–CALM module (AV-connection) further enhances the decay of activations in the output module.

After a response production, because all activations in the output and CALM module decay or are inhibited, the process of response production can repeat itself indefinitely, as long as input activations are provided to the word–CALM module. The output–CALM module combination can, therefore, be considered as a kind of parallel–serial converter. Such a converter is, of course, required for producing sequences of responses as is the case in free recall. We know of no other parallel–serial converter formulated fully in neural network terms that allows for performing free-recall tasks in a dynamic fashion. For example, in a model for free recall by Nolfi *et al.* (1990) entire 'sequences' are represented by a single output vector. This vector contains all recalled patterns as an unordered *set*, available as a whole, rather than as a *sequence* of subsequentially produced patterns.

General description of the simulations

All simulations with the ELAN-1 model proceed in three stages. In a first stage, old representations have to be formed in order to enable the model to perform implicit memory tasks. In a second stage, a subset of the already represented words is presented in a new study context. In a third stage, memory for the words is tested either implicitly (a word-completion task) or explicitly (a free-recall task).

First stage. The first stage of the simulations corresponds to pre-experimental learning. It simulates the acquisition of the kind of knowledge a subject possesses before taking part in an experiment. In real life, both implicit and explicit memory tasks are also dependent upon pre-experimental learning by the subjects (e.g. a subject has to be able to recognize a word). Therefore, before the simulation of the actual memory experiments can begin, artificial subjects have to be created by exposing the 'empty' network to its basic lexicon in a number of different environments or 'contexts'. Because in CALM modules the formation of representations is driven by a stochastic process (i.e. the E-node random activations), each presentation of the same series of words in contexts may lead to a different network. The differences among the trained networks may reflect individual variations between real subjects. (In these simulations, however, different artificial subjects were created by presenting the same set of words in different random orders to counterbalance ordering effects.)

In the first stage of the simulation, 20 different combinations of 10 word beginnings and 10 word endings were presented repeatedly under all but one of the seven contexts until sufficiently stable context and word representations had developed. (This makes $20 \times 6 = 120$ presentations.) With each of 10 word beginnings two alternative word endings were presented. Each word beginning and word ending was used twice in the basic word pool. The input nodes (one word beginning, one word ending and a context node) corresponding to an input pattern were given an activation of 1.0, whereas all other input nodes had zero activation throughout each learning trial. (The number of times words were presented and the pairing of words and contexts are described below in the actual simulation of the effect of word frequency on implicit and explicit memory.) A single learning trial lasted 100 iterations, which was more than sufficient for complete convergence in both CALM modules (see Section 3.2). Between presentations all activations (but not the weights) were initialized to zero (except for the A-nodes which were set to 0.2 and the output V-node which was set to 1.0 in order to prevent premature response productions). Because only 20 of the available 25 R-nodes in the word–CALM module can become occupied by word representations, five uncommitted R-nodes remained in this module. Occasionally one of the uncommitted nodes could win the competition instead of a committed node. Sometimes a word would also converge on a node that had been committed to another word. Such 'errors' were rather scarce and occurred only in about 1% of the cases after each word had been presented 16 times. The whole presentation sequence of creating the basic word pool in an initial network was repeated 12 times, so that 12 different, artificial subjects were created.

Second stage. In the second stage of the simulations, after initial training, an experimental list of words (a subset of all words learned) is presented to all artificial subjects with a new, not previously used, experimental context. The experimental words used in the second stage all had different word beginnings and endings selected from the total word list. These words were presented under the last remaining context not used in the first stage. Each experimental pattern was presented once for 100 iterations with an activation of 1.0 of the relevant input nodes. Between presentations of the 10 words the activations were initialized (see description of the first stage). A random order of presentation was determined before the simulation of the experiment, which was used throughout the second stage. The presentation of the experimental words was preceded by a single presentation (without word) of the experimental context during 100 iterations in order to obtain a classification of the experimental context before presentation of the words (pretest context learning).

Third stage. In the third stage, two test modes that correspond to the word-completion task and the free-recall task are simulated. The explicit *free-recall* task is simulated by activating the experimental context as an input pattern (see Figure 4.4). This can be compared with the instruction given to a subject to generate all words learned under the study context. In the context module the experimental context representation will become activated. This context, in turn, activates the words connected most strongly with it. Of these words, one will be selected by relaxation of the competition in the word module and this will lead to a response in the output module. Since the production of a response causes the resetting of all activations in the word module, the activation coming from the input context module may subsequently give rise to another response. This sequential response production process will continue as long as the context is available as an input pattern.

In the simulation of a *word-completion* task only the word beginning of a stored word is presented as input (see Figure 4.5). Of course, in this case no context is specified, which may be compared with the absence of a reference to the study context in implicit tasks. The active word-beginning node causes the activation of the nodes for all words with the same beginning. Words presented earlier in the experimental context will be connected somewhat more strongly with the stimulus and thus have a slightly better chance of being generated as a response than words not presented in the experimental context.

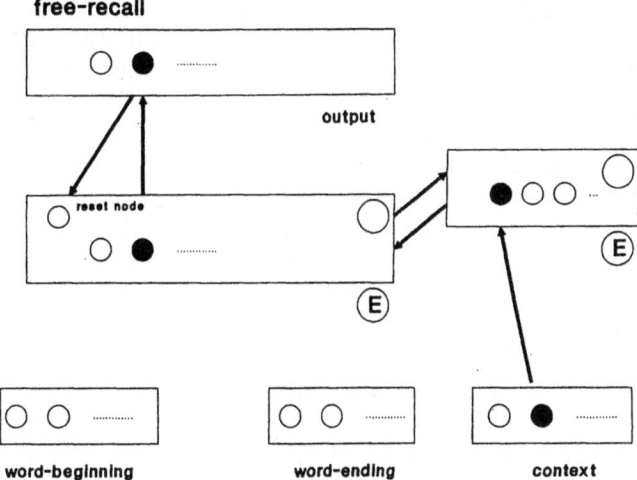

Figure 4.4 The ELAN-1 model with free-recall testing. Active nodes are denoted by full circles. With free recall the experimental context is provided as a cue, which leads to the activation of several words in the word–CALM that have been presented under the experimental context. A single word is selected by relaxation of the competition and output by the output module. After resetting both word– and context–CALMs the experimental context can activate another word.

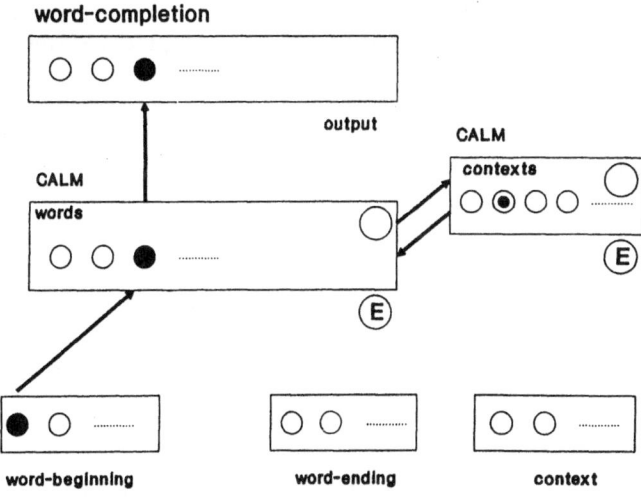

Figure 4.5 The ELAN-1 model with word-completion testing. Active nodes are denoted by full circles. A word beginning is given as cue and subsequently a full word is selected and output as a response.

Both the free-recall and the word-completion tests were applied before and after the learning of the experimental list for each artificial subject. In this way, the experimental performance of the simulated subjects can be compared with their base rate performance. Because the effects of the prelearning tests on the network were undone by restoring the artificial subjects in their pretest state, the experimental presentations that followed were not confounded by the test and, therefore, the testing before and after experimental presentation can be treated in the model as a within-subjects factor. This situation can, of course, never be obtained with real subjects, much as we might want it, because the first test would irreversibly change the subject and affect the performance in the second test.

In the free-recall task the experimental context was presented for 2000 iterations with activation value 1.0. The sequential response production by the combined output–CALM modules, which was described earlier on, allows for more than enough (over 40) attempts to recover words from the experimental list. Both with word completion and with free recall, learning (mainly simple activation learning) continues. The initialization of the activations of the nodes does not affect the weights, and the model, therefore, also shows learning during retrieval. Word completion involved the presentation of the word stems of the experimental words in a random order which differed from the order of experimental presentation. Each word stem was presented for 100 iterations with an activation of 1.0. After each presentation of 100 iterations all activations were initialized. Also with word completion, sequential responses were produced, but usually the same response was produced repeatedly in the series. The first response was always taken as the response produced by the artificial subject.

The parameters used here for CALM are those listed in Appendix A. All simulations were performed using the CALM Development System (discussed in Appendix B5).

4.2.4 The word-frequency effect

Experimental findings
The first kind of memory experiment to be simulated with this model has been performed both in our group (Phaf *et al.*, submitted) and elsewhere (MacLeod, 1989). In the experiment by Phaf *et al.* the effect of word frequency on implicit and explicit memory was tested. It was hypothesized that a single presentation would strengthen the representations of low-frequency words relatively more than those of high-frequency words (e.g. McClelland and Rumelhart, 1986c) due to their relative novelty. Consequently it was expected that presentation of low-frequency words would result in a larger beneficial effect in an implicit memory test than presentation of high-frequency words. Such an effect may be contrasted to the often reported finding that high-frequency words are better recalled (in a free-recall test) than low-

frequency words, especially if both types of words are presented in separate lists (Gregg, 1976; Hall, 1954). Because a word-completion task was used as an implicit memory test, words with a low spontaneous completion frequency (not to be confused with word frequency) were selected as target words. With such words an increase in the probability of completion due to earlier exposure can be detected more easily. The selection was based on norms derived from the spontaneous completion of 168 Dutch word stems by 200 subjects (Phaf and Wolters, in preparation). Target words had a normative completion rate of 0.03. From these words, two lists with either a low or high frequency of occurrence in the Dutch language were selected (less than 5, and more than 80 per million, respectively; Uit den Boogaart, 1975). Two groups of 20 subjects were presented with the list of either high- or low-frequency words. In the interval between presentation and free-recall test, one of two different tasks was introduced: either counting backwards or word completion. In the word-completion task the target word stems were mixed with distractor word stems which could not be completed to target words, so that there would be less of a chance that the subjects would become aware of the memory test character of the word-completion task. After performing the experiment, the subjects were asked whether they had been aware of the memory test character of the word-completion task. Those who had been aware (five subjects) were replaced by new subjects.

As was expected, and as illustrated in Figure 4.6, free recall (after counting backwards) was better for high-frequency words than for low-frequency words. Performance on the word-completion task revealed a substantial increase in the probability of completing word stems to the presented words,

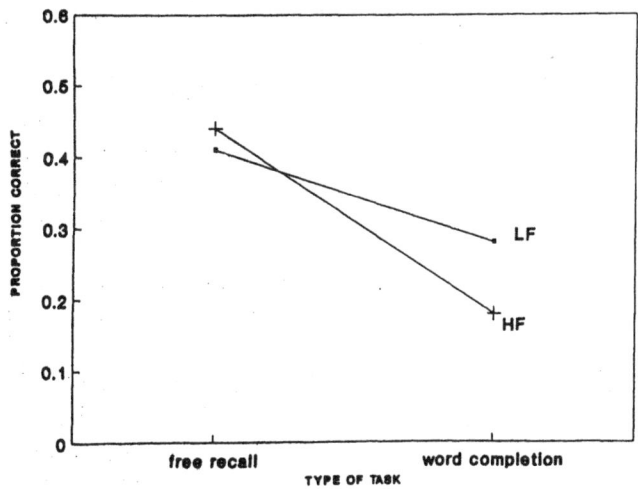

Figure 4.6 The interation between type of task (free recall or word completion) and word frequency (HF or LF) found in the experiment by Phaf *et al.* (submitted).

both when compared with the base rate of 0.03 and when compared with the distractor stems which had a similar base rate. Moreover, Figure 4.6 shows that the probability of completing word stems was larger for low-frequency words than for high-frequency words. A similar effect with a slightly different word-fragment completion task has been obtained by MacLeod (1989). The opposite effects of word frequency in word completion and free recall clearly reveal a significant dissociation between the implicit and the explicit test ($F(1,76) = 4.629$, $p < 0.05$).

Simulation of the word-frequency effect

In these simulations, the ELAN-1 model and general procedure discussed earlier were used. In this simulation the high- and low-frequency word representations were established by presenting 10 words twice as often (32 times) as the other 10 words (16 times) in the first training stage.

First stage. Assuming that word frequency is reflected in the number of contexts with which a word is associated, high-frequency words were presented eight times in four different contexts each (C1 to C4) and low-frequency words were presented eight times in only two contexts (C5 and C6). Each high-frequency word had one word beginning in common with a low-frequency word, but they always differed in word ending. The learning procedure was repeated 12 times (i.e. to create 12 artificial subjects). For each artificial subject a different random order of presentation was used with the restriction that two successive patterns would always have a different context. After these presentations the networks had learned a basic word pool consisting of 10 high-frequency and 10 low-frequency words. Before presentation of the experimental lists all artificial subjects were tested with word completion and free recall to determine the base rate for every artificial subject.

Second stage. In the second stage, 10 words, either all high frequency or all low frequency, were presented to each artificial subject in the experimental context (C7). Starting from the state of the network obtained in the first stage, each artificial subject was prepared in either of two different ways: a high- or a low-frequency list was presented. One could say that the artificial subjects were 'cloned', and that two identical 'clones' received different lists (see also Figure 4.9 later). In this approach, word frequency was, thus, also a within-subjects factor.

Third stage. In the test stage the 12 artificial subjects (i.e. 'cloned' pairs, each with the high-frequency and the low-frequency preparation) were confronted with both a free-recall test and a word-completion test. A successful recovery occurred whenever the cue presented (the context or word stem) would produce a target word in the output module. In order to determine whether activation of a particular output node signals an experimental word we have to know which word converges upon which output node. For this purpose the convergence data from the second stage were used. In the second stage, the 20 words presented to the subjects generally converged on the same nodes as in the first stage. A small number of classification errors occurred, however. Three artificial subjects had one node each with two words converging upon it. These words had either a word beginning or a word ending in common. In the test stage this node was taken as the correct classification for both words converging on that node.

Results: free recall. For each artificial subject a free-recall test consisted of running the model for 2000 iterations with a continuous activation of the input node representing the experimental context. This number of iterations was more than sufficient to ensure asymptotic levels of recall (see Figure 4.7).

The free-recall scores of the artificial subjects are shown in Table 4.1. In the simulation, average free-recall performance clearly increased as a consequence

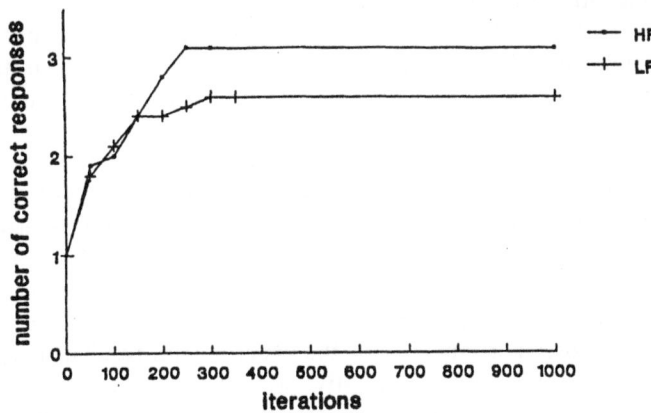

Figure 4.7 Cumulative free-recall curve for the first 1000 iterations in the simulation of experimental free recall. In the following 1000 iterations of the simulation no new words were recalled.

Table 4.1 Base rate and experimental free-recall performance (number of correctly recalled words out of a possible 10) for the 12 artificial subjects.

Art. subject	A	B	C	D	E	F	G	H	I	J	K	L	Average
High-frequency words:													
Base rate	2	1	0	0	0	0	0	0	0	0	0	0	0.25
Exp. recall	3	3	2	2	3	4	4	3	4	4	2	3	3.08
Low-frequency words:													
Base rate	2	0	0	2	0	0	0	0	0	0	0	0	0.33
Exp. recall	1	2	3	3	1	3	2	5	2	2	4	3	2.58

of experimental presentation. In principle, experimental recall may be taken for granted as absolute scores. However, since the results are derived from a very simple model with a strongly limited word pool, we also determined base rate free-recall scores, i.e. recall with experimental context as cue before the words have actually been presented with that context. This of course is something one would never do with real subjects. As can be seen from Table 4.1 base rate recall was very low. Corrected for base rate, high-frequency words are recalled better (2.83) than low-frequency words (2.25). So, just as with real subjects, high-frequency words are recalled better than low-frequency words, although the difference is not significant ($F(1,11) = 1.67$, n.s.).

Results: word completion. Word-completion performance also shows a clear increase as a result of experimental presentation (see Table 4.2). In this implicit task, low-frequency words benefited much more from the experimental presentation than did high-frequency words. The average increase in completion performance compared with base rate for the low-frequency words (4.33) was significantly greater than that for the high-frequency words (1.08) ($F(1,11) = 16.17$, $p < 0.005$). In the actual experiment a similar, but somewhat smaller, significant effect was found. The difference in the magnitude of the effect between experiment and simulation can be explained by the fact that in the experiment the base rates for high-frequency and low-frequency word completion were the same, whereas in the simulation they clearly differed. When high-frequency words have the same completion rate as low-frequency words, this is presumably caused by the presence of more or relatively strong competitors with the same word stem. In the simulation equalization of completion base rates can only be obtained by including more competitors to the same word stem. Although this manipulation would bring the results of the simulation in closer agreement with the experimental results, it would probably not change them in any qualitative way. The simulation results showed the same significant interaction between type of task and word frequency ($F(1,33) = 19.46$, $p < 0.001$) as the experimental results (see Figure 4.6). Figure 4.8 shows this interaction for the simulation.

Table 4.2 Base rate and experimental word-completion performance (number of correctly completed words out of a possible 10) for the 12 artificial subjects.

Art. subject	A	B	C	D	E	F	G	H	I	J	K	L	Average
High-frequency words:													
Base rate	9	6	5	3	5	7	7	5	5	5	5	6	5.67
Exp. compl.	9	8	5	8	5	7	7	7	6	6	6	7	6.75
Low-frequency words:													
Base rate	1	3	5	6	4	2	2	4	4	4	4	3	3.50
Exp. compl.	7	6	8	7	9	8	4	10	10	10	7	8	7.83

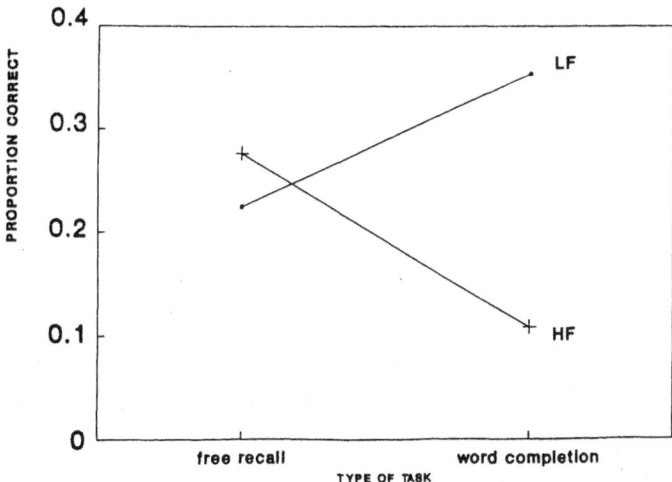

Figure 4.8 The interaction between word frequency (HF or LF words) and type of task (word completion or free recall) in the simulation. The sources are increases in performance (in terms of proportions correct) as a result of experimental presentation which have been obtained by subtracting the base rate performance from the experimental performance.

It may be noted that when high- and low-frequency words are presented within the same list, the frequency effect decreases or even vanishes in both explicit memory tasks (e.g. Gregg, 1976) and implicit memory tasks (Graf and Mandler, 1984, Exp. 2). To test whether the model would generate similar effects, the simulation was also run with lists consisting of mixed high- and low-frequency words. The simulation results with mixed lists also agreed well with the experimental findings. The details of this simulation are discussed in Phaf (1991) and in Phaf *et al.* (1990a).

The simulation method described here has some aspects that warrant extra attention. In the simulations, the experimental method is closely approximated by using artificial subjects. These subjects are provided with a 'personal history' by having them learn a background of high- and low-frequency words in different 'personal contexts'. Then they are presented with experimental lists in an 'experimental context', after which several memory tests are administered. The results may be analysed with the usual statistical techniques. In this way, the simulation method remains in close correspondence with the reality of psychological experiments. There is, however, one aspect that could never be found in reality. In the first and second stage, the artificial subjects are 'cloned', so that the performance of the cloned pairs on different tests can be compared directly. By using the 'cloning method' a within-subjects design becomes possible. Thus, after having acquired a 'personal background' in the first stage, an artificial subject AS is cloned. The two clones receive different experimental lists (high- and low-frequency words) to arrive at two clones

AS/HF and AS/LF. The cloning procedure is repeated in the third stage, where the two clones of a pair are cloned once more. These clones receive either a free-recall or a word-completion test. This results in a total of four clones for each artificial subject. Actually there are even more clones, because the cloning procedure is also used to derive base rate measures and convergence data. The cloning procedure is summarized in Figure 4.9.

Interpretation of the model's behaviour

The explanation of the small advantage of high-frequency words in the free-recall task is straightforward. When the experimental context cue is activated, all R-nodes in the context–CALM module will initially receive some activation. Because the high-frequency words in the word–CALM module are activated by more different contexts than the low-frequency words, they will have a small lead which gives them a slightly larger chance of winning the competition. A closer inspection of the weights from the context– to the word–CALM module clearly revealed that the total weight leading to a high-frequency word after experimental presentation was higher than the total weight leading to a low-frequency word after experimental presentation of the low-frequency word. In other words, a high-frequency word is more accessible by outside (environmental) information than a low-frequency word. When episodic knowledge is tested, the fact that this knowledge is embedded in a more elaborate network of associations proves to be an advantage.

The explanation for the word-completion effect in the model is more difficult. From the results in Table 4.2 one would expect that after experimental presentation the word-beginning-to-word connections of the low-frequency words have larger weights than those of the high-frequency words. Although *increases* of these connection weights (due to experimental presentation) were indeed larger for low-frequency than for high-frequency words, the average weight of these connections for the low-frequency words (0.472) remained below the average for the high-frequency words (0.475). The advantage for the low-frequency words seems to lie in higher connections of the

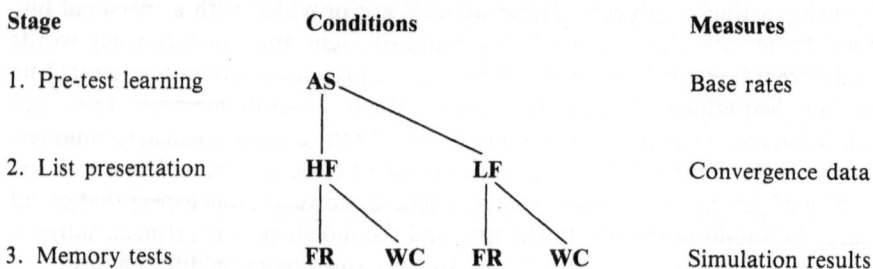

Figure 4.9 Summary of the design. In the first stage, artificial subjects are created. These are then 'cloned' to arrive at a within-subject design (AS = artificial subject, HF = high frequency, LF = low frequency, FR = free recall, WC = word completion).

word nodes to and from the context nodes. During presentation of the experimental list these connections are strengthened more for low-frequency than for high-frequency words. Also during presentation, through the bidirectional connections, activation of a word node will also lead to the activation of a number of context nodes and subsequent feedback to the word nodes. When a low-frequency word is presented, it will have a smaller initial activation than a high-frequency word and it will have a less clearcut advantage above other word nodes in the word–CALM module. During the competition phase, several contexts will become activated, so that the feedback from the context–CALM module will be more diffuse for the low-frequency words. As a result, somewhat more elaboration will take place with low-frequency words than with high-frequency words. Low-frequency words thus elicit somewhat more elaboration learning and this specifically enhances their representations, so that, when, in the third stage, a word beginning is presented the positive feedback from the context nodes will give more support to the low-frequency words than to the high-frequency words.

By showing that word-completion performance for high-frequency words is less affected by further presentation than low-frequency words the model illustrates on a microscopic scale how more general knowledge may develop from initial episodic storage through repeated presentation (e.g. Wolters, 1984). The general principle is that identity information embedded in a stationary network of many connections is affected less by later presentations than identity information that has not been embedded in this manner.

4.2.5 Simulating implicit and explicit memory in anterograde amnesia

In a recent book review in *Nature*, Marshall (1989) wondered whether connectionist models could simulate specific behavioural disorders. The second simulation might be an attempt to do just this. In this study we have simulated the finding that implicit memory is unimpaired in patients with anterograde amnesia, whereas these patients are severely impaired on explicit tasks like free recall and recognition (Warrington and Weiskrantz 1968, 1970; for a review see Shimamura, 1986). Recently, similar results have been obtained with fully anaesthetized surgical patients (Roorda-Hrdlicková *et al.*, 1990). After recovery from anaesthesia these patients showed no explicit recall, but performed well on an implicit memory test. On a category exemplar generation task (i.e. 'Mention some flowers.') subjects tended to generate words presented during anaesthesia.

We simulated anterograde amnesia in our model by artificially lesioning the noise generator that is driven by the novelty of a local pattern of activations in a module. This was done by setting the connection from the A- to the E-node equal to zero (see Figure 2.2), so that no noise could reach the representation nodes, and the learning rate remained at a constant low level.

In other words, we disabled the process of learning by elaboration in the model. Under these circumstances it was hypothesized that the network would no longer be able adequately to form new word-context representations. Consequently, free-recall performance for information stored after the lesioning should be low.

This artificial lesioning may somehow be analogous to the hippocampal damage responsible for anterograde amnesia. It has long been known that lesions of the medial zones of the temporal cortex generally give rise to severe disturbances of memory (Bekhterev, 1900; Grünthal, 1939). Since the 1950s it has been suggested that (bilateral) lesioning of the hippocampus causes anterograde amnesia (Milner, 1970; Mishkin, 1978; Scoville, 1954; Scoville and Milner, 1957; Penfield and Milner, 1958). Based on these and similar data from neurology we may assign the A/E-node in the CALM module a role comparable with some particular function of the hippocampus. There are also other reasons for suggesting that the A/E-nodes in the network play a role similar to the hippocampus in the human information processing system. For instance, Squire and Zola-Morgan (1988, p. 175) observe that: 'The output from hippocampus seems to be relayed to many areas of neocortex.... Moreover, these projections are largely reciprocated'. The same description would apply to the connections from a collection of E-nodes to other parts of our network model, if all E-nodes were put together in a single module. Also, the fact that the hippocampus is not itself implicated in the storage of the actual memories, because memories stored before the damage are generally not lost, is comparable with the role E-nodes play in our model. The E-nodes only play a role in the formation of memory traces and the integration of formerly unrelated representations in other parts of the system. It has to be noted, however, that there are numerous reports of synaptic changes and learning in the hippocampus (e.g. Kelso et al., 1986). In terms of our model these changes may relate to adaptivity of connections between E-nodes, which we have not modelled. The modification of such connections may play a role in the activation of many or all E-nodes by different kinds of external (arousing) factors (see e.g. Shors et al., 1989).

The procedure used for simulating anterograde amnesia was almost identical to the procedure used for simulating the previous experiment. The same artificial subjects obtained in the first stage of the simulations described above were used. The high- and low-frequency preparation was, therefore, maintained in this simulation. Amnesic artificial subjects were created by setting the AE weights to zero in both CALM modules before the beginning of the second, experimental, stage, so that neither CALM module could learn by elaboration. Otherwise the experimental learning and testing procedures were similar to the first simulation.

As can be seen in Table 4.3, word-completion performance is not affected by the lesion. Apparently, word representations stored before the lesions are not affected by the anterograde amnesia of the network. Free recall, however,

Table 4.3 Explicit and implicit memory performance for amnesic artificial subjects (AAS) lesioned before experimental presentation on free-recall and word-completion tasks. Experimental lists consisted of either high-frequency words or low-frequency words.

AAS	A	B	C	D	E	F	G	H	I	J	K	L	Average
Free-recall performance													
High-frequency words:													
Base rate	2	1	0	0	0	0	0	0	0	0	0	0	0.25
AAS	1	1	1	1	1	1	1	0	1	1	1	1	0.92
Low-frequency words:													
Base rate	2	0	0	2	0	0	0	0	0	0	0	0	0.33
AAS	1	2	1	1	1	1	1	0	1	1	1	1	1.00
Word-completion performance													
High-frequency words:													
Base rate	9	6	5	3	5	7	7	5	5	5	5	6	5.67
AAS	7	7	8	8	9	9	9	8	7	9	8	10	8.25
Low-frequency words:													
Base rate	1	3	5	6	4	2	2	4	4	4	4	3	3.50
AAS	7	7	7	7	9	7	7	6	10	8	6	8	7.42

is strongly affected by the lesion. Experimental presentation in this condition has almost no effect on free-recall performance.

An analysis of variance on the simulation results of both normal and amnesic artificial subjects showed a highly significant interaction between type of artificial subject (normal/amnesic) and type of test (free recall/word completion) ($F(1,11) = 46.35$, $p < 0.001$). Free recall of both high-frequency and low-frequency words is severely affected compared with normals, whereas word completion for low-frequency words is about the same and for high-frequency words even somewhat better (compare Figures 4.8 and 4.10).

The model shows a severe deterioration of explicit performance following learning after the lesion has been made. The formation of new associations between the experimental context and the words is clearly important for the execution of explicit tasks. Preventing elaboration learning disrupts the formation of such associations. The slight increase in word-completion performance may be due to the decrease of interference from new learning. Some indication for an increased implicit memory performance may also be found experimentally.

The advantage of low-frequency words over high-frequency words is preserved with word completion for the amnesic artificial subjects, though in absolute value completion performance for the high-frequency words is now higher. To our knowledge the effects of word frequency in a word-completion task have not yet been tested with real amnesic subjects. Although the model,

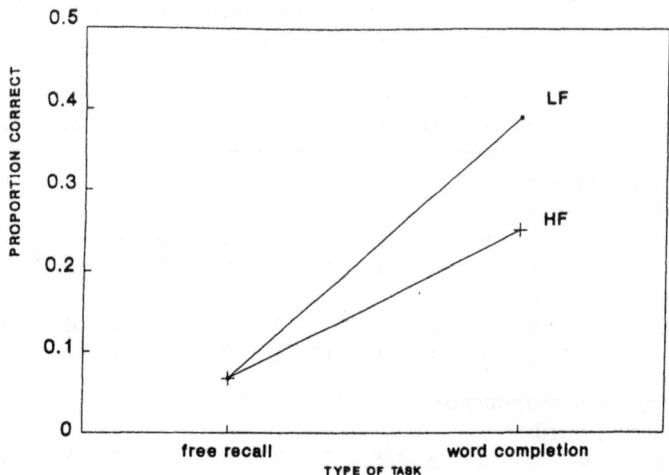

Figure 4.10 Increases (relative to the base rate) in proportions correct for the artificial subjects lesioned before experimental presentation (amnesic artificial subjects).

of course, only provides a strongly simplified account, it would be interesting to see if similar effects would really be found.

The essential point about this simulation is that it demonstrates that damaging a particular structure in the network affects only a certain kind of memory task (i.e. episodic memory tasks) without disturbing representations that have already been laid down in memory, and even allowing these to be strengthened still further. Anterograde amnesia is produced in the model by a processing deficit and not by a structural deficit in a particular storage system. Only the ability to build new episodic representations on the basis of more general, old representations has been lost by the amnesic artificial subjects.

It should be noted that the actual lesion performed produces the severest form of anterograde amnesia possible in the model, because the AE weights were set to zero. Milder forms may be found when a small value of the AE weight is retained. We have, in fact, found similar results in other simulations with less severe damage. With the severe damage almost no associations can be formed. With small values of the AE weight some ability to form new associations may be preserved. Such differences may, for instance, explain in terms of the model why amnesic patients with relatively mild memory disorders showed normal implicit memory for newly acquired associations between unrelated words, whereas severely amnesic patients did not show implicit memory for new associations (Graf and Schacter, 1985; Schacter and Graf, 1986; Shimamura and Squire, 1989).

The causes for amnesic performance in ELAN-1 are partly compatible with those assumed by McClelland and Rumelhart (1986b) in their distributed network model. McClelland and Rumelhart postulate that a reduction of increments in learning weights is responsible for anterograde amnesia and that it is not necessary to assume multiple memory systems. This explanation is similar to the reduced learning rate due to the absence of elaboration learning in ELAN-1. They call this the 'limited increment hypothesis'. Although their model shows some residual learning with amnesia, they do not succeed in showing completely spared performance in one task and severely impaired performance in another. In ELAN-1 we demonstrated such a dissociation by assuming two different learning processes, one of which is selectively impaired in amnesics.

4.2.6 Discussion of ELAN-1

Simulations with the ELAN-1 model show a number of similarities to the behaviour found in real implicit and explicit memory experiments. In the present model, word completion was chosen as the implicit test and free recall as the explicit test. Other tests, like threshold identification and recognition, can also be executed with this or a similar model. At present, threshold identification experiments have already been simulated with the ELAN-1 model (N.M. Mul, personal communication) by presenting words to the network for only a limited duration (in number of iterations) and with a small activation value. Presentation prior to testing leads to an increased number of correct responses by the model and reduced latencies under the threshold conditions. Simulations of recognition performance, probably, require the retrieval of the study context with the help of the word representation and the matching of the retrieved context to the reference context cue provided at test. The latter process would clearly only be possible with an extension of ELAN-1. An interesting prediction for this simulation would be that recognition of high-frequency words would suffer from competition with a large number of contexts, while recognition of low-frequency words would be less affected. An advantage for low-frequency words over high-frequency words is indeed often found with recognition (Gregg, 1976).

The activation–elaboration hypothesis implemented in CALM networks provides a strong basis for simulating implicit and explicit memory performance in a qualitative way. The network used here is a very simple one. The artificial subjects only have internal representations for 20 words and seven contexts. This means that in the experimental phase of the simulation half of the total set of words is presented. In a real subject, word and context representations are much more numerous and have a more intricate structure, including semantic relations between words (e.g. category information). Therefore, our artificial subjects are severely limited as a model for real subjects. None the less, the differential effects of word frequency are successfully simulated and can be explained in terms of the activation–elaboration hypothesis. Anterograde amnesic performance can be induced quite simply, and it is shown that it can be understood as an impairment of the elaboration process. Moreover, the model is also able to mimic the differential effects of retroactive interference in explicit and implicit memory tasks, which can be explained as a decrement in the effectiveness of the retrieval cue (described in Phaf (1991) and in Phaf *et al.* (1990a)).

We think the model shows that a multiple memory system explanation may not be necessary and that a multi-process view involving a continuum between learning with attention (elaboration learning) and learning without attention (activation learning) may be sufficient. More specifically, the anterograde amnesia result, which has often been used as an argument for multiple memory systems, can be very well understood in terms of different learning processes. Moreover, the latter view gives some specification of the role of attention in memory performance, whereas the former does not give any insight into what role attention may play. It should be noted that this also provides a significant extension to learning in connectionist models, because these models mostly possess only the ability to learn passively, which can be seen in the present context as activation learning. As we have argued before, however, to explain all dissociative phenomena in more detail an additional factor in the form of storage–retrieval compatibility may have to be included. This concession does not play a role in the multiple memory versus multi-process debate, however, since even some prominent multiple memory theorists seem to argue for adding this factor to their view (Tulving and Schacter, 1990).

Although qualitative agreement between simulations and experiment is good, quantitative agreement is somewhat poorer. As was already noted in the previous paragraph, this may be explained by the fact that the simulation differs from the experiment in many details. It is obvious that many factors that play a role in experiments with real subjects are not incorporated in the present model. At the moment, the model cannot represent category relationships and associations between words (i.e. interitem associations, see Raaijmakers and Shiffrin, 1980, 1981) other than through context representations. An additional CALM module connected to the word–CALM module may, however, represent word-category information. This extension, called ELAN-2, is currently being used for further simulations. Preliminary results with limited-size versions of ELAN-2 are encouraging. They will be described elsewhere.

Chapter 5

Pattern Recognition as a Practical Application

5.1 Approaches to pattern recognition

In the previous chapter, the application of CALM to modelling in psychology was discussed. In this chapter we will briefly discuss some practical applications of CALM, specifically, learning to recognize handwritten numerals. The model discussed in this chapter will be extended in the next, where genetic algorithms are applied to network design to reach much higher categorization and generalization scores. The model used in these two chapters is extremely limited, with a very coarse input grid and with virtually no preprocessing. Therefore, the preliminary results presented below fall short of the standards of real-world applications. They are presented only to illustrate the feasibility of the general approach to modelling and to outline various directions that could be followed to arrive at workable solutions. A recently commenced commercial project is aimed at developing real-world applications of CALM. In these models much more effort will be spent to make use of domain-specific knowledge in preprocessing and system design. To ensure a sufficient performance these models will be implemented in parallel hardware (see also Appendices B1–B5). The project will be primarily concerned with (autonomous) categorization of natural patterns. Before proceeding to discuss the model for handwritten character recognition, a few remarks will be made about two established approaches to pattern recognition and the way they are related to connectionism. We will, first, also briefly review some problems common to the recognition of both speech and handwritten characters.

Many research efforts have been invested in the recognition of auditory and visual patterns, because good solutions to these problems have great technological and commercial value. In 'classical' pattern recognition two main approaches may be distinguished: *statistical pattern recognition* (also called the decision-theoretic approach) and *syntactic pattern recognition* (see Krishnaiah and Kanal (1982) for an overview of both approaches, and see Fu (1974) and Gonzalez and Thomason (1978) for additional introductions to the latter approach). In statistical pattern recognition the aim is to assign patterns to one of a given set of pattern classes (i.e. recognition). Sometimes, an additional aim is to define the characteristics of some optimal set of classes (i.e. learning). In syntactic pattern recognition, recognition of the pattern is not limited to classification alone. These techniques also attempt to find structural descriptions of patterns. The syntactic or structural approach is inspired by linguistics, where a (natural) language is defined as a set of well-formed sentences, and well-formedness in turn is judged according to a grammar defining (i.e. generating) that language.

A successful parse of a sentence gives a structural description as a result. The structural descriptions of patterns are usually defined by some type of pattern grammar (e.g. a picture of a HOUSE is (i.e. consists of) a ROOF and WALLS; a WALL is a SQUARE, a DOOR, and WINDOWS, etc.). Different pattern classes may be defined by different grammars. If a grammar is unambiguous each element of the corresponding language has a unique structural interpretation. The structural decomposition of patterns that are coded as strings is often represented by a tree structure. Higher-dimensional pattern grammars exist, where instead of strings other mathematical objects are generated, such as graphs. Elements of a language defined by an ambiguous grammar may have more than one structural interpretation.

Structural descriptions may be important if patterns are very complex (e.g. visual scene analysis) or if the number of possible classes is very large (e.g. spoken sentences, visual textures, fingerprints). As Fu (1974, p. 2) argues, in these cases it is impractical to regard each description as defining a class, and 'the requirement of recognition can only be satisfied by a description for each pattern rather than by the simple task of classification'.

Some methods in pattern recognition incorporate aspects of the two approaches. Hidden Markov models (e.g. Rabiner and Juang, 1986), for example, may be viewed as a statistical approach, because they are primarily aimed at simple classification. But it can easily be shown that they form a subset of the set of stochastic regular languages, and that the time complexity of grammatical inference and string recognition are equal for stochastic regular grammars and hidden Markov models (Murre, 1987). Although the distinction between the two approaches is useful, we agree with Simon *et al.* (1982) that, perhaps, the justification for the distinction of the two approaches is more a matter of history than of strictly defined differences. In many ways, the two approaches complement each other.

Connectionist models are also an example of a method that combines elements of the two approaches. Many authors have pointed out similarities between certain types of neural networks and statistical techniques. For example, backpropagation can be viewed as a form of non-linear regression that is even suitable to perform a type of time-series analysis (Werbos, 1988). It may, therefore, be compared with standard techniques such as the well-known ARIMA models by Box and Jenkins (1970). Others have demonstrated similarities between certain types of Hebbian learning and principal component analysis (Oja, 1982). In Part I, it was shown how CALM can perform a clustering analysis. In short, connectionist models exhibit many characteristics of traditional statistical methods. The difference between neural networks and ordinary statistics may be more a matter of perspective than of inherent differences. There is one important aspect, however, that might be considered a distinguishing factor, namely the emphasis placed on *implementation* in neural networks: *many simple elements* running in parallel, and communicating simple signals through weighted channels. On the one hand, not all statistical techniques can be implemented in this particular form. On the other hand, many techniques for classification and recognition are unique to neural networks, such as self-organizing maps (Kohonen, 1989a, 1990) where the many-element representation of the maps is essential. The modular approach with CALM also seems to go beyond the ordinary clustering methods, in that it is possible to control the clustering and discrimination process by choosing different initial configurations of CALM modules. In this way, prior knowledge about the patterns to be analysed (i.e. learned and categorized) may be used to control the dynamics of the clustering process, as will be illustrated in the examples discussed below.

Connectionism also exhibits many of the characteristics of the syntactic approach to pattern recognition. Neural network models are, for example, able to learn sequences in a manner that resembles grammatical inference. Cleeremans *et al.* (1989), using a recurrent network of the type introduced by Elman (1988), demonstrated that such a network may learn to accept only sequences generated by some (stochastic) regular grammar. Long-distance sequential contingencies (i.e. of the type $\underline{a}ccc...ccc\underline{x}$ or $\underline{b}ccc...ccc\underline{y}$: a is always followed by x, and b by y) may cause problems in such networks if the many intervening elements (i.e. the cs) do not depend on the early information (i.e. the a or b). But Cleeremans *et al.* demonstrated that such long-distance contingencies can be encoded if the probabilities of the intervening elements are dependent on the early information. These simulations and others (Elman, 1989; Smith and Zipser, 1989) show that neural networks are able to perform grammatical inference, although the context-sensitive (i.e. non-symbolic) nature of representations in such systems may affect their behaviour in situations where long-distance contingencies are important (or deep embedding, etc.). Human subjects usually have trouble dealing with such dependencies as well, and it may well turn out that neural network implementations of syntactical knowledge result in models of greater psychological plausibility than some of the traditional psycholinguistic models. Neural networks are also able to deal with other types of syntactic structures, for example in music (Todd and Loy, 1991). Kohonen (1989b; see also Todd, 1989a) describes a non-connectionist learning algorithm, somewhat resembling an extended Markov model, that learns to generate music in the style of composers such as J.S. Bach. The system is fed with a large sample of the music, and then generates or 'composes' music in a style that is recognizably similar to the learned music. Although the texture of the music bears a remarkable resemblance to the training set there is little global structure (e.g. counterpoint) in the generated melodies. Kohonen (1989b) argues that such self-learning grammars should be viewed as 'associative memories of the second kind'. The idea is that sooner or later neural network research will catch up with these developments and find a way to implement such 'higher-order' procedures. Todd (1989b) has indeed developed a connectionist model that is able to learn regularities from a set of musical sequences and use these in a way comparable with Kohonen's method.

To our knowledge, no connectionist models have been described that produce something resembling a parse (i.e. a structural description) of a sequence. Such a system, a *neural parser*, would accept for example at the input side a string of discrete symbols, and at the output side produce a string representing a valid parse of the input string. The output string could represent a tree structure by using brackets (e.g. *aaacbbb* as input would give $(a(a(a(c)b)b)b)$ as output). The problem is not to develop a neural network building scheme that can do just this task for a given grammar, but to have the system *learn* to parse a string according to certain rules. Building schemes for neural networks that recognize a given context-free grammar have been proposed by some researchers, but such systems do not produce a parse (i.e. not as an output pattern sequence) and accept only strings of finite length. An early implementation of such a system by Fanty (1985), for example, was based on the well-known CYK parser (e.g. Hopcroft and Ullman, 1979). The most interesting aspect of this approach is that the parse is executed in parallel, working simultaneously on different parts of the input string, and reconciling bottom-up and top-down information. Fanty's parser can also dynamically add new grammatical rules (productions) if certain conditions prevail.

Although both the statistical and the syntactic approach to pattern recognition have been successfully applied to the recognition of natural patterns, such as speech and handwritten characters, there are several features of this task that make it seem exceptionally well suited for implementation in neural networks. Firstly, the recognition process involves many sources of evidence. It is a multi-constraint optimization task with a highly irregular structure. Such tasks can be easily mapped onto a neural network. Secondly, it is virtually impossible to program manually all the many rules that govern the recognition process. Therefore, most systems have taken recourse to some form of automatic rule induction or learning (e.g. the forward–backward algorithm for training hidden Markov models; see Rabiner and Juang, 1986). Neural networks have been shown to learn many difficult tasks and are, therefore, natural candidates for the task of learning to recognize speech and handwritten characters. Thirdly, neural networks are also able to interpolate missing information. Most network types can be trained to exhibit some form of content-addressable memory. From the discussion below it will become clear that this is an important asset of systems for speech recognition. Each handwritten text provides ample evidence of our ability to infer missing information in the visual domain. Indeed, even when correcting printed text – such as this one – we have to read with special attention, in order to override the automatic inference mechanism. The many printing errors that went undetected in the first draft of this book are proof of the ability and highly automatized character of this mechanism.

5.2 Sources of pattern variability

In Part II, we will be mainly concerned with the recognition of isolated, hand-written numerals, which is a relatively easy task compared with, for example, the recognition of continuous speech. It is instructive to review briefly some aspects of speech recognition, because it provides us with clear examples of the same difficulties we find in an attenuated form with isolated character recognition. Speech production is often taken as a paradigmatic task for motor production, and speech-related terms such as articulation and coarticulation have been generalized to include all motor skills (e.g. Jordan, 1990). Many of the problems for speech recognition also apply to systems aimed at recognizing cursive script (Schomaker, 1991) and − to a lesser extent − also to isolated character recognition.

Of all the recognition tasks routinely performed by humans, speech recognition is still one of the hardest tasks to be executed by a machine. No system has been devised that is able to do speaker-independent, continuous speech recognition with a large vocabulary in real time. Like handwritten characters, speech segments are extremely variable in appearance. Even isolated speech segments, produced by a single speaker in similar circumstances, may differ widely in timing, timbre, loudness, pitch and other characteristics. Speech segments may also greatly vary in the context of other speech segments. This is due to the nature of the articulation process. In the production of one speech segment we anticipate the next, whereas the previous segment may persevere in the production of the current segment. For example, if we say 'keep', the lips will usually be spread before saying the /k/ and may still be when saying the /p/. But if we say 'cup' no such spreading occurs. In other words, speech sounds are *coarticulated*. This makes it necessary for a recognition system to disentangle the mutual effects of the interacting speech segments. It may be argued from a psychological point of view that it is likely that knowledge about the articulation process may be an important factor for the high reliability of continuous speech recognition by humans. But this hypothesis remains to be tested.

Coarticulation is only one of the problems to be solved. Klatt (1979) mentions eight important problems that must be faced by a system for isolated word recognition (Klatt, more accurately, speaks about lexical hypothesis formation): (i) acoustic−phonetic non-invariance, (ii) phonetic segmentation, (iii) time normalization, (iv) talker normalization, (v) specification of lexical representations for optimal search, (vi) phonological recoding of word sequences in sentences, (vii) ambiguity caused by errors in the preliminary phonetic representations, and (viii) interpretation of prosodic cues to lexical identity. The human system is so skilled in solving these problems that if we hear a word in which a speech segment is totally obliterated by an intruding burst of noise, we may still hear an illusionary speech sound at that position (Warren, 1984). What we will 'hear' is the entire word plus an extraneous burst of noise at an uncertain location. This illustrates the strong mechanisms at work in the process of normalizing the variability inherent in continuous speech. We may still find those mechanisms operating at a very high level. Knowing the subject or intention of an utterance may result in 'hearing' words that may have been obliterated in the signal. Other high-level sources of knowledge may also be important, such as syntax, prosody and visual cues (cf. lip reading). In all of these cases, high-level sources of knowledge may induce an illusion of hearing sounds that are not physically present (cf. McGurk and MacDonald (1976) for the unlikely case of actually 'hearing' the effect of visual lip positions).

Like speech, the recognition of cursive script is also a very hard task, both for man and machine. It takes us, expert recognizers, perhaps 5 to 10 years before we can decipher the wide range of handwriting encountered in daily life. Depending on the aim of the recognition process, and on the nature of the data (e.g. isolated words, time-sampled handwriting movements, or cursive script) Klatt's eight problems apply equally to the recognition of writing. In addition to these specific problems, which mainly refer to stages in the recognition process, we may distinguish between two different sources of variability that must be reduced by the recognition process.

As spoken vowels, written characters exhibit a great variety. They may appear in different fonts, slants, thicknesses and so forth, and in many regional and individual writing styles. All of these factors may be put under the heading of *internal variability*. Internal variability causes a certain character to be defined by an entire set of patterns, rather than a single 'prototype' character (i.e. the set of all characters that can still be called *a*; see e.g. Hofstadter's (1985) discussion of fonts and styles, in particular his Figures 12-3 and 24-13). For processing visual patterns in a recognition system, some kind of preprocessing is needed. Elaborate systems may include stages like detection, recording, sampling, normalization, decomposition, segmentation and feature extraction. To some extent these operations can (and often *must*) be postponed to the recognition stage, which then of course becomes much more complicated. The resultant preprocessed patterns may, due to the nature of preprocessing, show distortions, such as rotation, translation, shrinking or expansion, and other affine transformations (e.g. caused by perspective distortions due to a non-centred viewpoint). These distortions by the system are also placed under the heading of *internal variability*, because like the 'natural variants' they show a high degree of regularity.

In addition to the 'regular' or affine distortions, at each stage of the processes of detection, recording, sampling, segmentation and feature extraction external noise may intrude. The detectors may have some faulty elements. The sampling may be too coarse for certain features, causing high-frequency noise.

Features may not be recognized. Broken lines may be wrongly connected by the system. All of these sources of variability in patterns may be called *external*, to distinguish them from internal variability of 'natural' variants and affine transformations.

Although visual patterns such as characters may exhibit a great variability, they may at the same time be encoded redundantly. A character of the ASCII code, for example, only needs 8 bits for representation. At an ordinary computer text screen such characters are typically represented by 35 or more bits. The same is true for natural, handwritten characters. The problem of character recognition, or any recognition problem for that matter, can thus be phrased as using the redundancy in patterns to filter out all sources of variability, and in this way to arrive at a unique categorization.

Not all sources of variability can be filtered out as easily by all neural networks. If a network is sufficiently 'expressive' (i.e. if it has enough nodes and weights, and the right initial architecture), it is usually relatively easy to deal with the internal variability of patterns, especially if the network architecture is constructed in a hierarchical fashion, where subpatterns (features) are extracted at a lower level and combined at a higher level. In this way, variant patterns may become encoded as combinations of different lower-level features. (Note that this aspect of neural networks somewhat resembles the hierarchical aspects of the syntactic approach to pattern recognition (Fu, 1974).) Also, the external variability caused by, for example, superimposed noise on the patterns causes little problem because of the associative nature of most neural networks. The main problem hides in the affine transformations that must be executed on the pattern in order to normalize it sufficiently. Another major problem is segmentation (i.e. isolating a pattern from a background of patterns).

The two most successful neural network models for handwritten character recognition that do not rely on elaborate preprocessing (in contrast to, for example, the Nestor system) are the neocognitron by Fukushima (1980, 1988; Fukushima and Miyake, 1982; Fukushima *et al.*, 1983; see also the review by Roberts and Li, 1988) and a model developed at AT&T Bell Laboratories (e.g. LeCun *et al.*, 1990a). Both use a modular architecture where different parts of the model specialize in solving subproblems. Like CALM networks, the neocognitron uses inhibitory and excitatory neurons bound together in a hierarchically structured, modular system. Learning is based on 'winner-take-all competition' with a variant of the Hebb rule developed by Fukushima. The neocognitron has many layers, each of which adds a small step towards invariant classification. This makes it virtually unnecessary to do elaborate preprocessing of the input patterns, but it also makes the model more complicated, harder to comprehend, and perhaps more difficult to implement than other models for pattern recognition. Nevertheless, it is able to learn in an unsupervised manner to recognize patterns reliably. The neocognitron also uses selective attention sequentially to (segment and) recognize simultaneously presented patterns. It still is one of the very few models that is able to do this.

The approach by LeCun *et al.* can also be called modular, in that they train several small networks to deal with subsets of the problem and then combine the result (see also Guyon *et al.*, 1989). Of particular interest is their method of achieving translation invariance. The input grid is scanned by an array of detectors each with a limited input field. An extra restriction (also described in Rumelhart *et al.*, 1986) forces weights at the same relative position in the input field of a detector to be equal for all detectors at that position. In this way, the detectors are trained on features independent of position in the input grid. This method is sometimes called 'spread'. A side effect is that the number of modifiable weights is strongly reduced. It can be shown that if a network with a reduced number of weights learns a certain task, it has a much greater chance of generalizing its behaviour beyond the training set than a more complex neural network (see also the discussion in Part III). Thus, the method of LeCun *et al.* also increases generalization performance. A disadvantage is that the method is not biologically plausible, and the training stage may be hard to implement in hardware as the forcing of equal weights would require either many connections (hardwired) or a detector would have to scan different positions in the grid sequentially. The network discussed in the next section has more in common with Fukushima's neocognitron than with LeCun *et al.*'s model. It will require, however, a lot more work before it makes any sense to compare its results with these models. As remarked at the beginning of this chapter, its inclusion here is for illustration purposes only. In the next chapter we will sketch several ways to improve its performance.

5.3 A small network that learns handwritten numerals

The network used for learning handwritten numerals consists of five CALM modules. The output (i.e. final categorization) is derived from a top module of size 15. The top module has bilateral connections to four smaller CALM modules, each of size 4 (see Figure 5.1). These four CALM modules get input from four 3×3 (overlapping) quadrants in a 5×5 input grid. Input between the grid and the four CALM modules is not direct. Activations in a quadrant are first led through a layer of nodes (one node for each unit in a quadrant) that have weak mutually inhibitory connections (shown as modules I1, I2, ... in Figures 5.1 and 5.2). This induces some 'sharpening' of the patterns in each quadrant. Each of the modules I1, I2, ... was fully connected to a corresponding CALM module (shown as C1, C2, ... in Figure 5.1).

Digits were written in the grid by means of a mouse and a computer screen. The percentage blackness in each square on the grid was translated into an input activation between zero and one. In the first simulation phases, 10 prototype numerals (i.e. a single set of handwritten digits 0, ..., 9) were presented repeatedly to the corresponding nodes in I1, I2, I3 and I4 (see Figures 5.1 and 5.2). The prototypes were presented in the fixed order 0, 1, ..., 9 until they were perfectly and consistently discriminated by the top module (C5 in Figure 5.1).

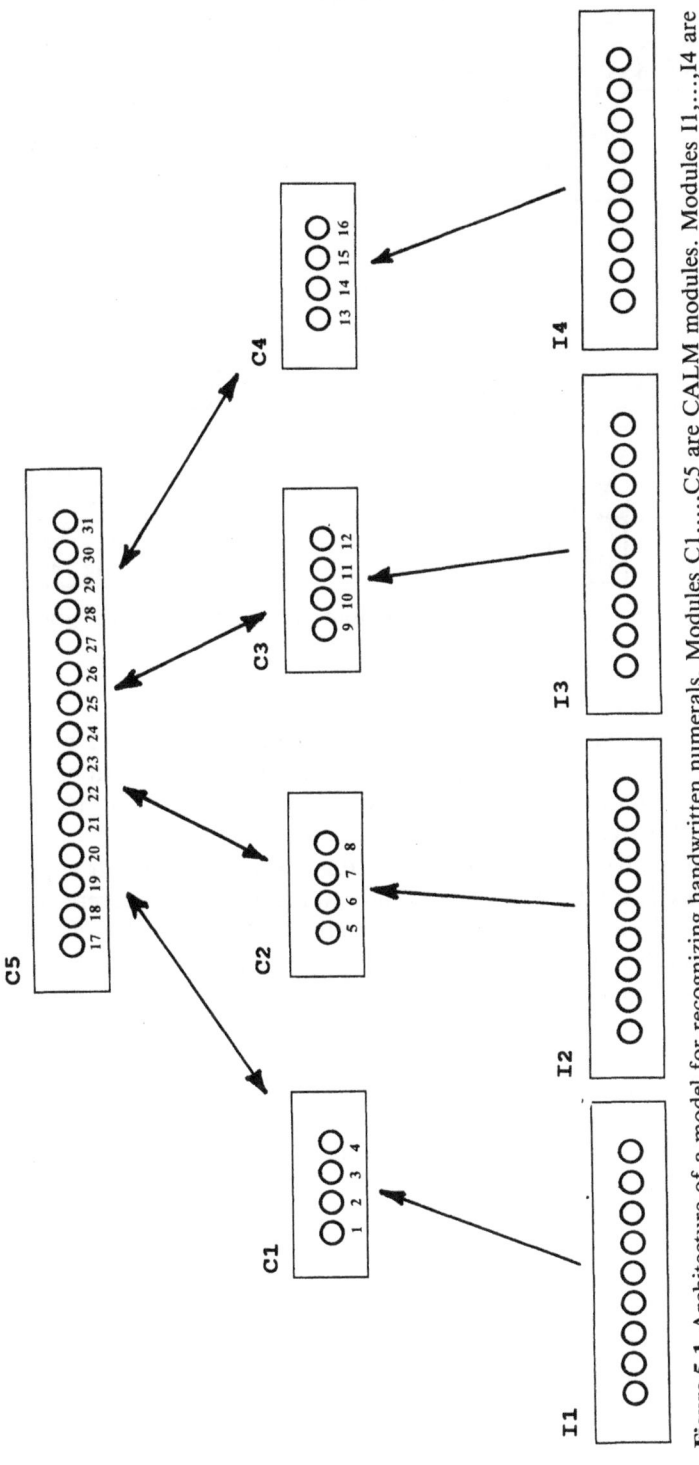

Figure 5.1 Architecture of a model for recognizing handwritten numerals. Modules C1,...,C5 are CALM modules. Modules I1,...,I4 are input modules (also see Figure 5.2).

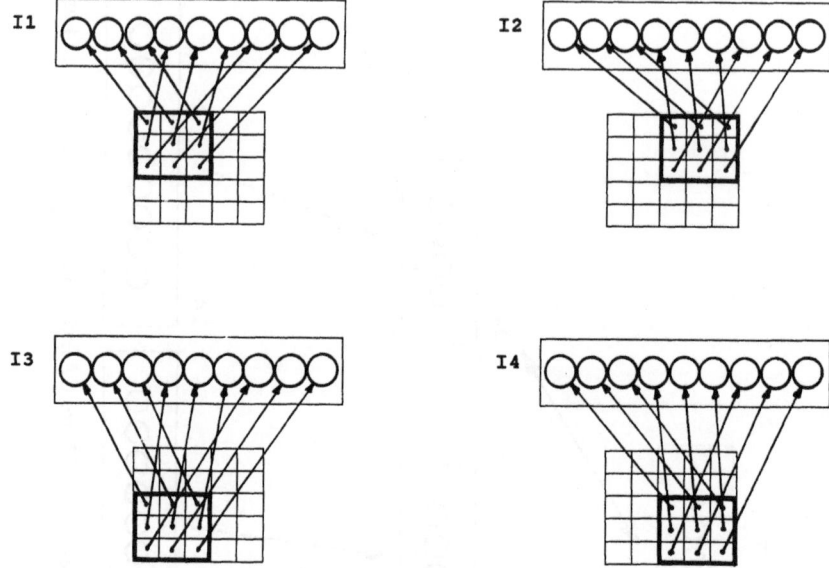

Figure 5.2 Connections from the four quadrants in the input grid to the input modules (see Figure 5.1).

Learning was unsupervised (i.e. no categories were forced by an external teacher). Presentation of a single pattern lasted for 50 iterations. Before each presentation, the model's activations were reset to zero. After 10 presentations of the 10 patterns a stable set of categories had evolved.

The receptive fields that emerged after learning in the four smaller CALM modules C1, C2, C3 and C4 are shown in Figure 5.3. The receptive fields were calculated by a linear approximation to the CALM activation rule (see Happel (1990) for details). Receptive fields for the top-level module C5, calculated by a similar approximation, are shown in Figure 5.4. The method of linear approximation of receptive fields may be criticized as too crude for large models. In such cases, a method developed by Kindermann and Linden (Linden and Kindermann, 1989; Kindermann and Linden, 1990) may be used that is based on an 'inverse' form of backpropagation. For the small network discussed here, the linear method seems satisfactory. As can be seen from Figure 5.3, the receptive fields of nodes in one module tend to be orthogonal. The receptive fields of the nodes in the top module partly show the original prototype with which the network was trained. These receptive fields show a large central cross caused by overlapping quadrants, which cause information in the central axes to be emphasized. Ignoring these cross effects, in the top-level receptive fields it can be seen with some difficulty which pattern is represented at each node. Of course, the 5×5 input grid was quite coarse, and this may obscure some of the details, so that even a human observer may encounter difficulties when trying to recognize such patterns.

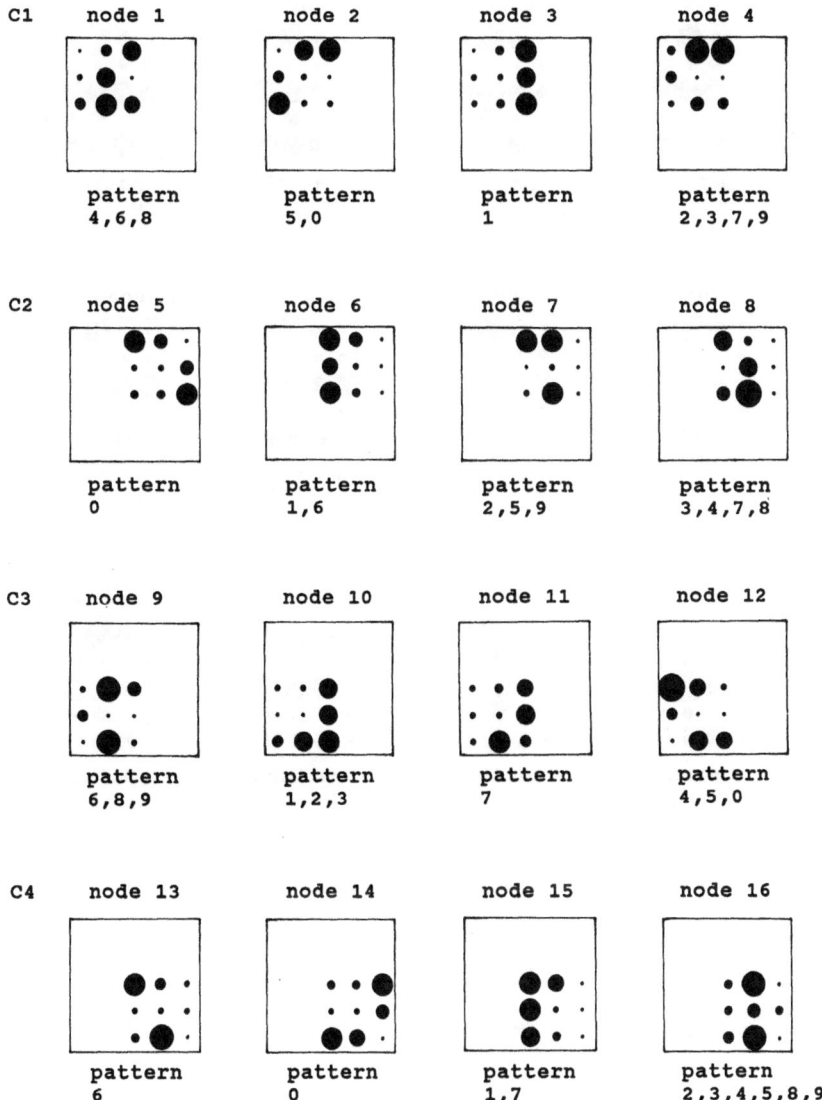

Figure 5.3 Receptive fields of R-nodes in CALM modules C1,...,C4 (see Figures 5.1 and 5.2). The size of the full circles indicates the sensitivity of each node to input at the corresponding points in each quadrant.

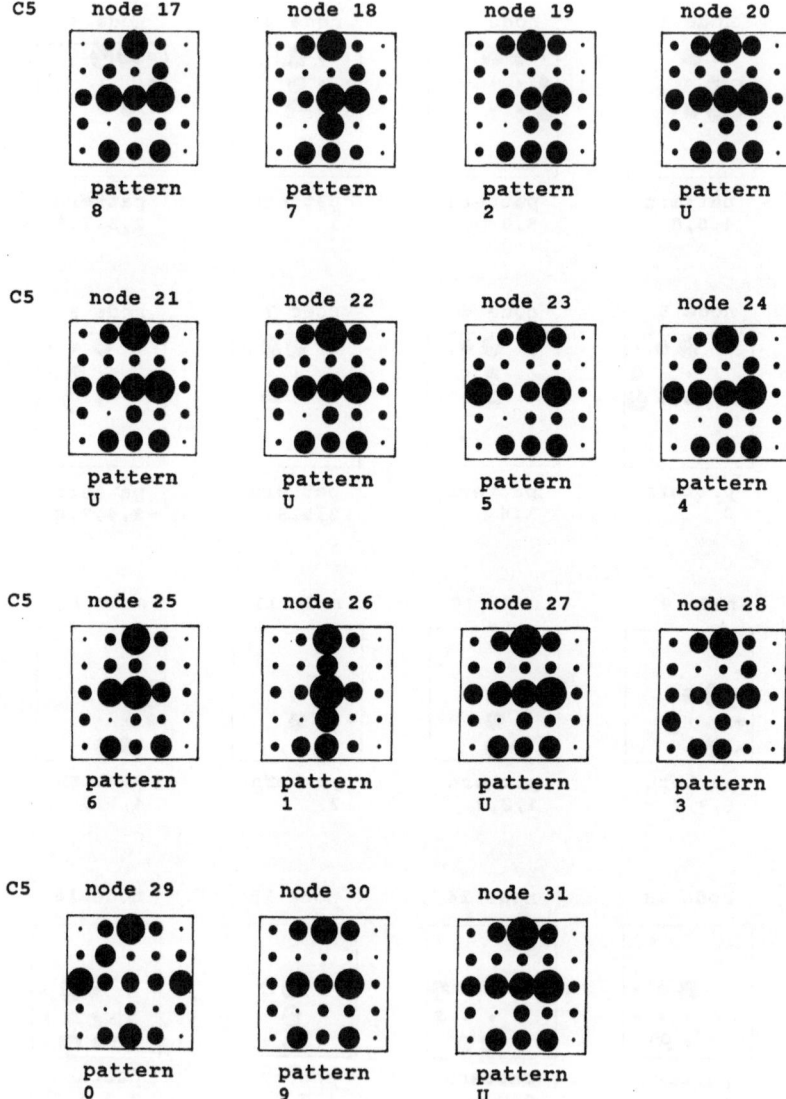

Figure 5.4 Receptive fields of the R-nodes in the top-level CALM module (see Figure 5.1). The size of the full circles indicates the sensitivity of each node to input at the corresponding grid points. U means uncommitted node.

Although the model could rapidly learn the prototype patterns in an unsupervised manner, the learning results did not generalize well to other handwritings. Of a set of 100 patterns (the numerals 0, 1, ..., 9 in 10 different handwritings) only 38.1% was categorized correctly (Happel *et al.*, 1990). The low generalization performance is most probably due to the coarse input grid and lack of preprocessing. Although these elements of the model can easily be improved, finding an optimal recognizer was not the primary objective here. We merely wanted to illustrate how patterns can be recognized by a CALM module. Many practical questions regarding the choice of the network remain unanswered. The solution of such problems is the subject of another research project. In the next chapter, however, we will address some possible solutions that are suggested by using genetic algorithms to optimize network parameters and topology. Genetic algorithms may not only provide efficient design methods: some deep similarities between neural networks and genetic algorithms suggest that computing methods gleaned from nature may themselves bear on more general laws, most of them still undiscovered.

Chapter 6

Genetic Algorithms: Modularity, Learning and Network Design

6.1 A brief introduction to genetic algorithms

In this chapter a computing method is introduced that is based on a biological metaphor of evolution: genetic algorithms. We will briefly review its principles and discuss some of its characteristics. Among others, it will be argued that modularity plays an important role with genetic algorithms as well, although in a somewhat disguised form. We will also discuss some aspects of the relation between learning and evolution. Then, some simulation studies that combine genetic algorithms with neural networks are briefly reviewed. Finally, we will apply genetic algorithms to the design of modular neural networks. In particular, we will investigate to what extent they can be used to improve the results of the model for handwritten numerals of the previous chapter.

Goldberg (1989, p. 1) defines genetic algorithms as 'search algorithms based on the mechanics of natural selection and natural genetics'. The process of evolution is used as a 'real-world model' that serves as a source of ideas for solving practical and theoretical problems in modelling and optimization. Just as neural networks use a 'brain metaphor', genetic algorithms rely on an 'evolution metaphor'. The general use of biological metaphors in modelling has strongly increased in the past decade. In a recent article, Valdés (1991) has coined the term *biocomputing* to refer to a field that, in addition to neural networks and genetic algorithms, also includes reiterated affine transformations (Barnsley, 1988), fractal systems (Mandelbrot, 1982), fuzzy logic (Kandel and Lee, 1979; Zadeh, 1987), chaos theory (Gleick, 1987), simulated annealing (Kirkpatrick *et al.*, 1983), and the study of complex systems (Prigogine and Stengers, 1984). The successes of nature in solving many problems that are very difficult for the traditional approaches have led researchers into studying the solutions of the biological example. In various abstractions and formalizations, biological systems have been theoretically and empirically proven to provide robust solutions to these hard problems.

The genetic algorithm is an example of a search algorithm that uses random choice as a tool to guide a highly exploitative search through a (coded) parameter space (Goldberg, 1989). This combination of a noise source that drives a search process structured by some selection mechanism also lies at the heart of the CALM algorithm. As we have pointed out in Part I, several other neural networks possess this general property (e.g. the Boltzmann machine, see Ackley *et al.*, 1985). Neural networks and genetic algorithms also share this property with simulated annealing (Davis, 1987). Another correspondence between genetic algorithms and the CALM approach to modelling is the importance of modules or building blocks. Although the role of the building block in genetic algorithms is quite different from the module in CALM networks, we will see that it also shares some unexpected similarities.

Genetic algorithms were introduced by John Holland from the University of Michigan. A classic reference is Holland (1975), but for this section much use has been made of the recent book by David Goldberg (1989). The interested reader is referred to this excellent introduction, which includes not only all the basics of genetic algorithms, but also a description of computer software (in Pascal) and a seemingly exhaustive overview of applications of genetic algorithms to a wide range of problems. Genetic algorithms have been applied to, among others, optimizing the layout of VLSI circuits (Davis and Smith, 1985), optimizing the connections in communication networks (Davis and Coombs, 1987), simulating the evolution of populations of microorganisms (Rosenberg, 1967), and the selection and creation of feature detectors for pattern recognition systems (Chang *et al.*, 1990; Englander, 1985). In his book *The Blind Watchmaker*, Richard Dawkins (1986) presents an amusing and instructive simulation in which he aims at letting trees evolve, specified by some genetic code. Dawkins himself determines the fitness of the trees (which are drawn on a computer screen). To his own surprise, he discovers that the system not only grows trees, but is also able to produce insects, faces, and other interesting figures. These novel shapes evolve within a few generations, apparently demonstrating the 'quantum character of phylogenesis' (Feistel and Ebeling, 1989). Radically new forms may appear very suddenly, within a few generations. As will be described below, genetic algorithms can be combined with neural networks to form even more powerful adaptive systems, suitable for the solution of many practical problems. But first let us very briefly review some of the basics of genetic algorithms.

Goldberg (1989, p. 7) mentions four differences between genetic algorithms and other search procedures:

1. Genetic algorithms work with a *coded* parameter set.
2. They search from a population of points, rather than from a single point.
3. They only use directly available information, usually a unidimensional objective (fitness) function.
4. They use probabilistic transition rules, instead of deterministic rules.

As with neural networks, the terms used for genetic algorithms differ from their biological counterparts. A biological organism is (partly) determined by its genetic code: genes with certain values on the chromosomes. An organism develops on the basis of the genetic code. The relation between the genetic code and the developing organism may be strongly non-linear: small changes in a gene may – but need not – result in great changes in the organism. An organism in a population has a certain chance to live and reproduce, which is usually called its fitness. Fit organisms live, others die. Or from a slightly different perspective: fit genes live, others die (see also Dawkins (1976) for a convincing defence of the 'gene perspective').

Genetic algorithms work with *strings* of (usually) binary features. A string *codes a possible solution to a problem*. Suppose we have a problem with only one parameter x. We want to find the optimal value for x, which is known to remain within the range $[0, 255]$. The parameter x might be coded as a conventional 8 bit binary string, for example $01001101 = 64 + 8 + 4 + 1 = 77$. Changing the second bit from 1 to 0 results in a decrease of 64 in parameter x, namely from 77 to 13. Changing the last bit diminishes the parameter by only one, resulting in a value of 76. Changing a single feature may, thus, affect the coded parameter very strongly or hardly at all. More complicated encoding may be constructed where the effect of changing a bit at a given position in a string is made dependent on the values of other bits in the string. This is the situation usually found in nature, where the effect of each gene is thought to be dependent on a few other 'conditional genes'. The number of 'conditional genes' may not be too large. It can be shown that, if each gene is dependent on too many other genes, the 'decoding process' may become unstable (Kauffman, 1991). This implies that the organism's early development would be severely hampered.

Genetic algorithms work with populations of strings, so we could imagine having 10 strings available, each with different feature (gene) values. As in nature, fit strings have a large chance of being reproduced. Evaluation of the fitness of a string can happen in many different ways. The method is blind to this evaluation; it is only interested in the result, a single number, often chosen between zero ('not worth living') and one ('optimal fitness'). The evaluation method can be denoted as a fitness function $f(x)$. In our example, the fitness function could be a transparent mathematically expressed relation such as $f(x) = x^2$, but also some parameter in a real-world problem could be used, such as $f(x)$ is 'the inverse of the air resistance of a car as a function of x, the angle of the front window in degrees'. In the first case, it would suffice simply to calculate f on a computer (or by hand). In the latter case, it would be necessary to build first a car according to the specification of the coded parameter, and then test it for air resistance. Although the latter method is much more expensive and time consuming, the genetic algorithm works exactly the same. The fitness value of a string is used to decide which strings will reproduce.

From the population of strings, two are randomly selected for reproduction, with the fit strings in the population having a higher chance of being selected. The general reproduction process is governed by so-called *genetic operators*: *selection*, *crossover*, *mutation* and *inversion*. They affect the fitness of the population. Of course, we want the genetic operators to *increase* the fitness of the total population, so they must be carefully chosen. Mutation randomly 'flips' a single bit in a string. Inclusion of the mutation operator is not necessary in most cases, but it may improve the performance by 'suggesting' now and then a new partial solution. For most problems mutation is not the most efficient genetic operator and the chance of mutation is usually chosen very low. Crossover is far more efficient in most cases. Crossover takes two strings and a string cutpoint (a position in the string, randomly generated). It cuts both strings at the given cutpoint and 'glues' the resulting pieces together, but only after the two pieces have been switched. For example, suppose we have two strings of 8 bits *xxxxxxxx* and *yyyyyyyy*, and a cutpoint 5. Cutting both strings at position 5 and interchanging the pieces gives two new strings *xxxxxyyy* and *yyyyyxxx*. The inversion operator takes one string and two cutpoints. It detaches the piece of string between the two cutpoints, turns the detached string around and 'glues' the pieces together. For example, applying inversion to an 8 bit string $x_1x_2x_3x_4x_5x_6x_7x_8$, with cutpoints 3 and 6, gives $x_1x_2x_6x_5x_4x_3x_7x_8$ (note that the order has been changed). In practice, crossover is used in most genetic algorithms, inversion is applied less often.

Randomness may enter in the genetic algorithm at various points. The initial population is randomly generated. Deciding which operator to apply to which string(s) is a stochastic process. Positions (i.e. for mutation) and cutpoints (i.e. for crossover and inversion) are randomly generated. This randomness drives a search process through the search space defined by the fitness function. Although the driving force is stochastic, the result is not a pure random search. The search proceeds by *selection* through fitness criteria and by *structured information exchange* through crossover or inversion. This combination gives genetic algorithms much of their power.

The above remark can be stated more precisely by using the concepts *schema* and *building block*. A schema (Holland, 1968, 1975) is a template string that defines a set of strings that have certain values at positions in common. As an example, 11*****0 is a schema defining the set of all strings that have two ones in the first positions, and a zero in the last position. The '*' is a wildcard or 'don't care' symbol; it fits either '0' or '1'. Suppose the fitness function is $f(x) = x^2$, then the strings described by this particular template string all have a fitness of at least (binary) 11000000^2 or $(128 + 64)^2 = 192^2 = 36864$. (Here, we do not limit the fitness value to the range zero to one.) The template 00*****0, on the other hand, contains only strings with a fitness equal to or lower than (binary) 00111110^2 or $(2 + 4 + 8 + 16 + 32)^2 = 62^2 = 3844$. We may, thus, associate a fitness value (or value range) with certain schemata.

Good solutions to a problem are coded by strings consisting of highly fit schemata. This is not surprising, of course, because highly fit schemata are better reproduced than schemata with a low fitness value. Crossover, however, has a tendency to cut long schemata in half, thus decreasing their chance of reproduction. Goldberg (1989), therefore, introduces what he calls *building blocks*: schemata of short defining length. We might also call them local schemata, or modules. Sufficiently short building blocks are left alone by crossover. As is shown in Goldberg (1989), fit building blocks are propagated from generation to generation, where the number of highly fit building blocks shows an exponential increase. It can also be shown that, per generation, about n^3 schemata are effectively evaluated for fitness. Because only n function evaluations take place, each involving several implicit schema evaluations, we may speak of a form of *implicit parallelism* in genetic processing. In the next section we will compare further the concept of a genetic building block with the neural network module.

6.2 Modules as partial solutions

We have already mentioned the similar roles of noise and selection in genetic algorithms and neural networks. Several other interesting similarities exist. The building block may find a parallel in the module concept that lies at the basis of this book. Building blocks represent pieces of 'knowledge that works'. The hypothesis is that the combination of a number of fit building blocks (i.e. partial solutions) will lead to a good approximate overall solution. Rather than starting from scratch at each generation, a genetic algorithm selects a few good building blocks and tries to recombine them (by crossover). In doing so, it retains partial solutions, trying to find better combinations and occasionally creating a novel partial solution (when cutting a building block during crossover or through mutation).

An interesting illustration of the modularity or building block hypothesis is a small study by Koza and Keane (1990). They applied genetic algorithms to two control tasks: cart centring and broom balancing. The former task involves moving a cart to a preassigned position. The latter task has also been solved by learning neural networks (e.g. Barto *et al.*, 1983). At each iteration, a controlling system is provided with the current state variables of the cart (position and velocity) or the broom (position, velocity, angle and angular velocity). The system can either apply a *constant* force in the left or right direction, or do nothing. The system acts according to a predefined *strategy*. It is the task of the genetic algorithm to find an optimal strategy. Strategies are formulated as LISP S-expressions (i.e. rooted point-labelled trees in the plane); they are hierarchical expressions (trees), recursively composed of a set of atoms such as the absolute function (ABS), greater than function (GT), multiplication (∗), and the state variables. The genetic operator crossover was redefined as the interchange of subexpressions (i.e. subtrees or subroutines), rather than substrings. From an initial population of 300 random strategies, for both tasks optimal or near-optimal strategies were found within five generations (near-optimal strategies performed within 1% of the analytically derived optimal solution). These results provide a direct example of the plausibility of the modularity thesis, even if *module* is interpreted as a software subroutine (LISP expression). Search methods based on this thesis may therefore be expected to perform well.

For complicated real-world problems any background knowledge available must be used to arrive at a workable solution. Modularly constructed systems are necessary to tackle these problems, because only they allow subproblems to be isolated and solved separately. With genetic algorithms, when defining the parameter coding of a certain problem, the researcher may exploit his or her knowledge of the problem by putting related (coded) parameters closely together in the string. It would not, for instance, make sense to break up a binary code (as in our example above) into two parts, positioning the first part at the beginning of a long string and the second part at the end. If the genetic algorithm could find a good parameter value, it would be bound to be destroyed by crossover, cutting the 'highly fit parameter schema' into two parts. (There are other ways to get around this problem, for example the use of two-point crossover. We will not discuss these methods here.) Similarly, we have argued in Part I that knowledge of the problem may be used in designing neural networks, by processing distinct aspects of a problem by different modules, the results of which are only combined at a higher level.

The thought behind both Goldberg's 'building block hypothesis' and the 'modularity thesis' advocated in this book is that replacement of a partial solution by a better one will in general give rise to better overall solutions. In particular, it is hypothesized that it will only very rarely lead to a worse solution. We may call this the hypothesis of 'continuity of solution'. If the quality of solution can be expressed by a single quantity, and if the partial solutions are represented as values in some multi-dimensional parameter space, then we expect the solution surface to be smooth: the best solutions are surrounded by good solutions, not by the worst. The latter case, where the best solutions are surrounded by the worst, can be called the 'needle in the haystack' problem, and its solution is very hard by any conceivable method. For extreme 'needle in the haystack' problems, no systematic method would perform essentially better than random search. The reason for this is that the performance of a particular combination of partial solutions would provide no clues as to the proximity or direction of a similarly (e.g. slightly better) performing combination.

The conclusion is that neural networks and genetic algorithms share some important basic properties. They have also been applied to the same problems, such as finding optimal feature detectors or solving the travelling salesman problem. A number of studies have appeared, comparing the performance of genetic algorithms with neural network and simulated annealing on NP-complete problems (e.g. Bounds, 1987; Brady, 1985; Davis, 1987; Spears and De Jong, 1990). To our knowledge, these studies have not yet resulted in a general theory or a set of guidelines indicating the strengths and weaknesses of either of the methods in a particular application.

6.3 The interaction of evolution and learning

It could be argued that genetic algorithms can be used broadly to guide learning in connectionist networks (Whitley and Hanson, 1989). Genetic algorithms are relatively insensitive to local minima and quick at finding approximate solutions, but they are typically slow in fine-tuning these solutions. Neural networks that use gradient descent learning, such as backpropagation, may perhaps be characterized as more sensitive to local minima and better at fine-tuning. A combination of both approaches could thus enhance overall performance. As was argued extensively in Part I, constraints on network topology (e.g. modular structure, or number and size of layers) form a major determinant of learning characteristics, such as learning speed and quality of solution. Specification of the topology by a genetic algorithm, therefore, seems to be the most likely approach: *genetic algorithms guide learning by determining learning constraints*. This can be likened to 'coarse-grained programming' of neural networks (see Appendix B5). By setting the appropriate weight values, learning imparts a finer structure on a neural network coarsely lined out by a genetic algorithm. As will be shown below, this combination of methods may perhaps also find application in the solution of practical design problems.

The performance of a genetic algorithm is strongly dependent on its evaluation or fitness function. The approach above suggests that a solution (i.e. a neural network topology) be evaluated according to its learning performance: neural networks that are able to *learn* a given task quickly are considered fitter than their slower competitors. As has been argued by some researchers (e.g. Hinton and Nowlan, 1987; Belew, 1989), *in this approach learning guides evolution*. Hinton and Nowlan (1987) illustrate this somewhat counterintuitive fact by an example where, in a neural network, L connections with binary weights (0 or 1) must be specified by a genetic algorithm. As fitness function they deliberately choose one that assigns non-zero fitness only to the perfect solution (i.e. all L connections right), and zero fitness to all $2^L - 1$ remaining solutions. This function turns the problem into a 'needle in the haystack' problem (see the discussion above).

A primitive form of learning is introduced to alleviate the 'needle quality'. The genetic algorithm no longer determines all weights, but some weights are left unspecified. These weights may be modified by the network. The 'topology' specified by the algorithm is thus limited to (i) the values of specified weights and (ii) the relative number of modifiable weights. A given network is evaluated for learning performance by counting the number of guesses g it takes to arrive at the perfect solution. At each guess, the network must reset all of its modifiable weights to a new binary value. The network is only permitted G guesses. The fitness function is made inversely proportional to the counted number of guesses: many guesses (high g) give a low fitness. If no perfect solution has been found at all within G trials, minimum fitness is assigned. Maximum fitness is achieved only if all weights are correct before any guessing has taken place, in other words when all weights are determined by the genetic algorithm and no weights are modifiable.

Adding learning to the original algorithm strongly increases the 'continuity of solution' of the problem. If the network has many connections, and if it is only permitted few guessing trials, the effect of learning will at first be negligible. It may take many generations before a network is generated with a fitness value above the minimum. But as soon as this happens, there is a fair chance that this network's offspring also has an above-minimum fitness value. The algorithm is now able to retain some of its knowledge in the following generation, and, furthermore, this knowledge tends to increase at every new generation, because networks will need fewer and fewer guessing trials to arrive at the perfect solution. Once the algorithm hits upon an above-minimum fitness value, it will converge to maximum fitness within a few generations. This may be expected to happen sooner if the number of weights L in the networks is lower and if the number of guessing trials permitted G is higher. What happens here may be pictured as a 'needle in a hole', where the 'hole' or basin of attraction is formed by learning. *Without* learning we are glancing from above trying to discern the needle standing up (which will look very tiny). *With* learning we move over the surface until we reach the hole which allows us to find the needle quickly. Learning has turned the 'needle in the haystack problem' into a 'hole in the haystack problem'.

The conclusion may be that the influence of learning and evolution is bidirectional. At the level of the individual network, learning is constrained by evolution, which has specified its initial architecture. At a global level, evolution is directed by learning, which creates a basin of attraction through modification of the fitness evaluation, as was illustrated in the example of Hinton and Nowlan (1987). Or, in other words: *in ontogenesis evolution guides learning; in phylogenesis learning guides evolution*. Recent theorizing in the philosophy of biology also stresses the interrelatedness of local and global levels in evolution and adaption (i.e. learning). For instance, Depew and Weber (1989; see also Weber *et al.*, 1989) have presented a well-argued proposal to integrate the global, ecological perspective and the local, developmental perspective in neo-Darwinist evolutionary theory. Their arguments are philosophical, and the theory is not applied to specific instances. Recent simulation studies of artificial ecosystems do provide nice illustrations of their viewpoint.

Ackley and Littman (1990; more fully described in Ackley and Littman, in prep.) studied evolution in an artificial world populated with agents, carnivores, trees, plants and walls. They were specifically concerned with the interrelation between learning and evolution. An important conclusion was that learning may provide a decisive evolutionary advantage. Another study by MacLennan (1990) gave similar results. He studied the evolution of finite-state machines that are able to communicate. As in the former study, learning greatly enhanced survival. Another conclusion by MacLennan (1990) was that communication may evolve in a population of simple machines that are physically able to sense and modify a shared environment, and for which there is a selective pressure on cooperative behaviour. Communication further increased the average fitness and the speed of fitness increase of the population. Both studies illustrate that learning and communication are favoured by evolution. Genes enhancing learning in organisms have a higher chance of survival than less adaptive variants. Genes that enable organisms to communicate tend to have high survival value as well.

These and other relations between learning and evolution may lead us to a combined genetic–connectionist approach that may not only contribute towards further development of neo-Darwinist evolutionary theory, but also provide us with additional common ground for biology and psychology to study problems in learning and development.

6.4 Genetic algorithms and neural networks

Recently, researchers have started to investigate whether the above sketched approach, with genetic algorithms determining neural network topology, is useful in practice (e.g. Dodd, 1990; Maricic and Nikolov, 1990; Miller *et al.*, 1989; Whitley and Bogart, 1990; Whitley and Hanson, 1989; Whitley and Starkweather, 1990). Others have taken a more direct approach, where not only the topology and learning parameters are specified by the genetic algorithm, but also all weight values and thresholds (De Garis, 1990a,b; Harp *et al.*, 1989). In many of these studies, a comparison is also made with traditional search methods. From these comparisons it appears that the combination of neural networks and genetic algorithms holds much promise for practical applications. Because it is impossible to review all of this research here, let us limit ourselves to one typical example of each approach.

De Garis (1990a) uses a genetic algorithm to determine the weights in a neural network. He calls this type of network, which is trained or rather programmed (fine-grained programming, see Appendix B5), by a genetic algorithm GenNets. The network described controls a two-legged, walking stick-figure. Whenever one of the feet becomes lower than the other, it is said to be 'on the ground'. By alternating its 'foot on the ground', and by moving its leg joints to and fro, the figure is able to 'walk'. As input the network accepts four joint angles and four angle velocities, and as output it produces four angle accelerations. The network consists of 12 nodes: eight input nodes and four output nodes (no hidden nodes). It is fully self-connected and uses a symmetrical sigmoid activation function. The distance covered (to the right) by the stick-figure within a given time is taken as the fitness criterion. Only two genetic operators are used: selection and mutation. De Garis (1990a) argues that the weights in his neural network are so strongly interdependent that interchanging parts of the network (i.e. as a result of crossover) is detrimental to their performance. This appears to contradict the modularity thesis. Elsewhere (De Garis, 1990b), however, he states that when a modular approach to network generation is taken, network modules can be independently 'programmed' by a genetic algorithm. In such a case, one would expect that interchanging parts of a single module (here, a tightly interconnected subnetwork) would similarly impair its performance.

Several series of experiments show that De Garis' system is indeed able to produce walking stick-figures, but the first attempts resulted in 'most unlifelike' walking behaviour. Additional constraints were, therefore, imposed on the motion of stick legs. This did not produce satisfactory results either: the evolved figures made one big step, and remained stuck with overstretched legs. Finally, a single network was first programmed for the task of 'stepping'. This network was then used as a parent to produce networks that could take several steps in one direction, which was the original aim.

Dodd (1990) also underwrites the thesis that for certain problem types *structured* neural networks are necessary. He proposes to use genetic algorithms to optimize generalization ability and compactness of networks by imposing constraints on connectivity and network parameters. Rather than determining the network structure in full, including all weight values, only the architecture (topology and network parameters) is specified. He argues that 'for problem types where the mechanism of data production is well understood, human engineering of suitable network structure is often possible. Where we do not have this knowledge it is desirable to make use of an automatic method for network optimization.' (Dodd, 1990, p. 694). This viewpoint fully coincides with that taken in this book. Two tasks are discussed, both of which involve a pattern recognition problem. The networks use 'spread' to achieve translation invariance, a method introduced by LeCun *et al.* (1990a; see also Section 5.3). Feature detectors have a fixed-size input window that moves over the input pattern. In this way, the weight values found are not dependent on any particular position on the pattern. The parameters of the architecture include height and width of the replicated weight pattern, the horizontal and vertical increment of this substructure, learning rate, and a set of initial weights.

In the first part of the paper, Dodd compares the performance of genetic algorithms with a pseudo-gradient descent method. Fitness is taken to be inversely proportional to the residual error and network complexity (details are not given in the brief paper). The task is a line orientation problem. The results show that the gradient descent method finds networks that perform very well, but that their performance is highly dependent on the initial weights used. When the performance is re-evaluated with different sets of initial weights (the network has 100 weights) it drops to less than half its original value. By contrast, the genetic method, though it performs slightly less well, is not dependent on initial weight values, and it generalizes very well. Dodd (1990) also briefly describes a real-world application, where a genetic algorithm is used for the specification of parameters for a neural network that recognizes dolphin sounds. In 30 generations, the method produces networks (with 5000 weights) that perform better than human-engineered networks.

The conclusions of these two studies are that genetic algorithms and neural networks may be fruitfully combined for application to difficult control and recognition problems. The studies also show that genetic algorithms may produce unexpected, and sometimes unintended, answers to the problem the researcher is trying to solve, and that the results achieved (i.e. the structure of the final parameter set) can be very hard to interpret. In the following section, two experiments in the application of genetic algorithms to designing CALM networks will be described. Our experiences reinforce the above conclusions.

6.5 Designing modular networks with genetic algorithms

As part of a national research initiative, a project was started in 1990, sponsored by the (Dutch) Neural Network Foundation, that is aimed at developing methods and guidelines for the design of neural networks for practical applications. The work described in this section forms part of this project, which is not limited to the research of CALM networks.[*] Two experiments in genetic engineering will be described, aimed at exploring ways to improve the model for character recognition in Section 5.3. In Experiment 1, a set of feature detectors is 'programmed' by a genetic algorithm. In Experiment 2, various architectures are tested in a genetic search for optimal recognition performance.

The genetic algorithm used in the two experiments below was the 'static population model' or GENITOR algorithm developed by Whitley (1989). This model has a number of advantages. It uses 'rank-based selection', where the selection probability is proportional to the rank of individuals in the population. This circumvents the problem of fitness scaling. And it uses 'one-at-a-time selection'. A new solution replaces the worst member of the population rather than its parents. In this way, the best solutions always remain in the population, and the average and best fitness values increase monotonically. In the experiments, the population consisted of 100 CALM network configurations. Initially, a random population was generated. In all simulations, the genetic operators used were two-point inversion ($p = 1.0$), two-point crossover ($p = 1.0$) and mutation ($p = 0.005$).

6.5.1 Experiment 1. 'Programming' feature detectors

In Section 5.3, a small neural network for the recognition of handwritten numerals was described. We will use this network as a starting point for improvement by genetic algorithms. In Experiment 1, a genetic algorithm is used to search for (i) optimal parameter values for the CALM learning and

---

[*] Bart Happel is the primary researcher on this project, and it is he who has carried out the simulations reported here (see Happel and Murre, 1992, for additional results).

activation rule, and (ii) optimal weights for a module of 16 (non-learning) feature detectors. As in Section 5.3, we were primarily concerned with exploring the possibilities and difficulties of the combined approach to pattern recognition, rather than with developing a better recognizer. As the results indicate, this could well be the subject of another project that is directed more to applications.

The model of Experiment 1 resembled that of Section 5.3, with the exception that between the 5×5 input grid and the CALM output module (of size 10) a module with a single layer of 16 feature nodes was inserted. Each of these feature nodes received input from the entire input grid. The weight values of the incoming connections were non-learning, and determined by the genetic algorithm. The activation rule for the feature nodes was similar to normal CALM nodes. The 'programmed' weights were encoded in the bit string by 4 bits, so that they could only take on 16 different values. A genetic value, G, in the range 0–15 was recoded to $G \times Inter/100$, where $Inter$ is the initial value of the learning interweights. In Table 6.1, the value of $Inter$ was established as 0.848, a rather high starting value that was also used by the learning weights in the upper module. With this value for $Inter$, the feature weights were in the range [0.0, 0.1272]. Initially, random values were generated in this range for these weights. The CALM weights and parameters were initialized to the values given in Appendix A.

There is one other deviation from the 'standard CALM module'. Extra

Table 6.1 Weight and parameter values found in Experiment 1. The values may be compared with those of Appendix A.

	Description	Value
Weights:		
Up	Connects R-node to its matched V-node	1.749
Down	Connects V-node to its matched R-node	−0.810
Cross	Connects V-node to all non-matched R-nodes	−8.896
Flat	Interconnects V-nodes	−1.135
High	Connects V-nodes to A-node	−10.600
Low	Connects R-nodes to A-node	0.322
AE	Connects A-node to E-node	1.005
Strange (or ER)	Connects E-node to R-nodes	0.175
Inter	Initial value of learning weights	0.848
VE	Connects V-nodes to E-node (only used in Exp. 2)	−5.000
Parameters:		
k	Decay of activation in an iteration	0.027
K	Maximum value of learning weights	1.156
L	Learning competition	2.854
d	Base rate of learning	0.130
$W_{\mu E}$	Virtual weight between E-node and learning rate	0.050

VE-connections were included in the CALM module between the V-nodes and the E-node. This was done because we wanted to check whether this added control of the arousal system would result in better performance.

As in Section 5.3, the task was to learn an input set of 100 handwritten digits (10 sets of 10 handwritten digits, written by 10 different subjects). The same set as in Section 5.3 was used. For a single fitness evaluation of a network, the entire input set was presented five times to the network. Each presentation lasted for 60 iterations, sufficient for convergence and learning to take place (see Part I). Fitness evaluation was based on the percentage correct score after learning. The goal was to maximize recognition for the input set.

After 3000 recombinations and network evaluations the fittest network in the population achieved 97% correct recognition (i.e. only three of the hundred digits wrong). When the fittest network was tested on 50 new digits (five sets of 10 digits), the results generalized to 69% correct recognition, which was substantially better than the results found in Section 5.3 (38.1% generalization).

Although this learning task is in a sense atypical, because the feature weights were 'programmed' rather than learned, we may still wonder what parameter values the algorithm had found. The weight and parameter values are listed in Table 6.1 and may be compared with those listed in Appendix A. It is interesting that the algorithm has chosen a very low activation decay ($k = 0.027$, compared with $k = 0.05$ in Appendix A). The algorithm has not completely shut off the arousal system, although the low high-weight value of -10.6 indicates that the A-node will not have been active very long at a

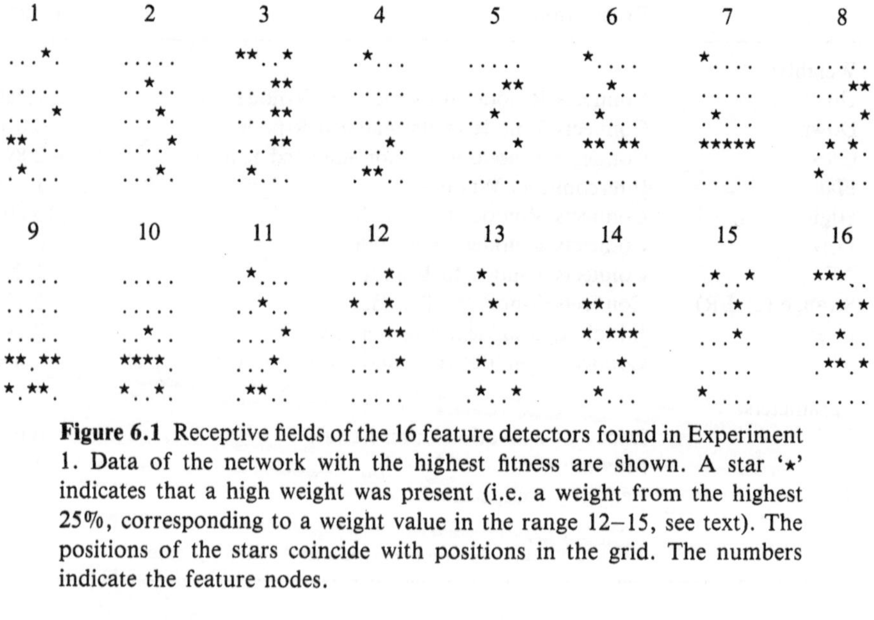

```
   1          2          3          4          5          6          7          8

...*.      .....      **..*      .*...      .....      *....      *....      .....
.....      ..*..      ...**      .....      ...**      ..*..      .....      ...**
...*      ...*.      ...**      .....      ..*..      .*...      .*...      ....*
**...      ....*      ...**      ...*.      ....*      **.**      *****      .*.*.
.*...      ...*.      .*...      .**..      .....      .....      .....      *....

   9         10         11         12         13         14         15         16

.....      .....      .*...      ...*.      .*...      .....      .*..*      ***..
.....      .....      ..*..      *..*.      ..*..      **...      ...*.      ..*.*
.....      ..*..      ...*      ...**      .....      *.***      ...*.      ..*..
**.**      ****.      ...*.      ....*      .*...      ...*.      .....      **.*
*.**.      *..*.      .**..      .....      .*..*      .*...      *....      .....
```

Figure 6.1 Receptive fields of the 16 feature detectors found in Experiment 1. Data of the network with the highest fitness are shown. A star '★' indicates that a high weight was present (i.e. a weight from the highest 25%, corresponding to a weight value in the range 12–15, see text). The positions of the stars coincide with positions in the grid. The numbers indicate the feature nodes.

presentation. This is also evident from the low VE-weight of -5.0. The non-zero low weight implies that in the initial phase of a pattern presentation the arousal system may become active for a short time. In this type of simulation, with a fixed pattern presentation order, the arousal is only necessary to break the symmetry. In other simulations, where unexpected patterns may intervene, the arousal system remained fully operative (B.L.M. Happel, personal communication).

The receptive fields of the feature detectors of the fittest network are schematically represented in Figure 6.1. The receptive fields broadly resemble those found in Section 5.3, in that they tend to be orthogonal. They do not directly reflect recognizable digits, but rather seem to encode more general features. The weight values learned by the CALM module are not shown here. They remained mostly in a narrow range of 0.10–0.25. These low-weight values agree with the analysis of Part I on equilibrium weight values, which leads us to expect low values if many inputs are present (even if the individual input activations are low).

6.5.2 Experiment 2. Exploring architectures

In this experiment the objective was to investigate to what extent genetic algorithms may be used to design modular neural networks. The genetic algorithm could change (i) the CALM parameters (see Experiment 1) and (ii) the global network architecture. The architecture of the network models used in these simulations consisted of a 5×5 input grid and a 10 R-node CALM module (called output module here) for categorizing the handwritten digits 0 to 9. In between the grid and the output module, a network of additional CALM modules was present. The *modular structure* of this network was determined by the genetic algorithm. A constraint was that the *total* number of R-nodes in the intermediate CALM modules was not to exceed 15. The number of modules was not a constraint, but of course if there were, say, five modules, they would have only three R-nodes on average. Also, modules were allowed that remained unconnected with either the input grid or the output module (unused modules). There were no constraints on the connections. Bidirectional connections were allowed, as well as direct connections from the grid to the output module. As always, if two modules A and B were connected, this implied that all nodes in A were connected to all nodes in B.

The training scheme was similar to that of Experiment 1 (100 handwritten digits were presented five times for 60 iterations). A difference with Experiment 1 was that the VE-connections were not used. The fitness was based on percentage recognition of the training set. The initial population consisted of 100 networks with random architectures and the CALM parameter set (Appendix A). The algorithm was run for 3000 recombinations.

The 10 fittest networks all reached recognition scores of 92%. The algorithm did not find any network with a higher score, so that we may wonder whether this is the maximum score that can be attained with CALM on this training set. The fact that this result is worse than that of Experiment 1 (97%) may be explained by the larger search space for the 16 feature detectors. A feature detector 'programmed' by a genetic algorithm may assign any combination of weights, but if it is trained by some learning rule, the constraints in this rule limit the combinations that can be found (e.g. see the analysis of equilibrium weights in Part I). The networks were also tested for generalization to a set of 50 digits in five new handwritings. Of the networks that achieved 92% recognition, the best performance was 74% generalization (no other network did better).

The architectures of the 10 fittest networks included only three non-trivial architectures. Seven 'architectures' were found in which the input grid was connected directly to the output grid, without any intermediate modules. These networks reached generalization scores from 54% to 74%. The differences in generalization can be attributed to differences in the parameter sets found for these networks. Interestingly also the other three networks maintained a direct connection between the input grid and the output module, supplemented by an indirect route that involved some kind of rough categorization in CALM modules of size 2, 3 or 4 (see Happel and Murre, 1992, for a more detailed account of these and other simulations).

In the early generations more complex networks were formed with approximately five intermediate CALM modules. Many of these also reached 92% recognition. The fittest, complex networks tended to organize themselves with architectures that had three streams: (i) direct connections between input and output, (ii) indirect connections via a stream of CALM modules with four R-nodes, and (iii) indirect connections via a stream of CALM modules with two R-nodes (see Happel and Murre, 1992). This architecture bears some resemblance to that of the visual system (e.g. Zeki and Shipp, 1988). We may speculate that the different level of coarseness of analysis by the magnocellular and parvocellular pathways has arisen because it represents an efficient method for recognition in modular neural networks.

From these experiments, it is concluded that genetic algorithms may be applied to the automatic design of modular networks, and that in this way better results may be achieved for practical tasks. The use of genetic algorithms also has a heuristic value. Networks may be generated that suggest certain principles of design. Of course, a more rigorous analysis is then necessary to verify the 'suggestions' by genetic algorithms. As argued above, a more mature science of neural networks and genetic algorithms may eventually lead to a better understanding of the laws governing the emergence of species and their capacities for learning and adaption. Simulation studies are a necessary testing ground for these more encompassing theories.

6.5.3 Implementation of the genetic algorithm

All simulations described were performed with the GENLIB, a software package in C, developed at the Leiden Connectionist Group by Bart Happel. The GENLIB was linked to the CALMLIB (see Appendix B5) using small application-specific programs. Simulations were run in parallel on about 10 VAX workstations (of various types) in our network using a very simple, but highly effective, self-synchronizing communication scheme (a similar system is also mentioned by Dodd (1990)). In this scheme, all network descriptions in the current population are stored in a single, globally accessible population file. Since with these applications evaluation of a network takes much more time than generating a new offspring, a process consults the population file only once in a while. After a process finishes the evaluation of a specified number of networks, it locks the population file, compares the fitnesses recorded with the fitnesses found, and inserts the newly found solutions, if their fitnesses are sufficiently high (thereby erasing the worst-fit networks in the file). After unlocking the file, the new solutions become available for all active processes in the network. These may lock the file, select two parents, and generate a new network description, after which they unlock the population file (without any further changes) and start to evaluate the network performance. Other processes, running on different machines, may at any time try to access the population file and update its contents. Although this method does not exclude the possibility that two processes are evaluating the same network, the chances of this happening are low as long as the population size is much larger than the number of processes. In practice, it is hardly worth complicating the communication scheme for this reason. We are currently working on a version of the system running on a transputer network of 32 or 64 transputers.

A new project is also being started to transform the GENLIB into a fully-fledged 'genetic simulator' that will be called MetaGen. This system will be developed along the lines of the MetaNet Network Environment (see Appendix B5). The two systems will be integrated to the point where they can easily communicate with one another. The idea is that genetic algorithms can be created in MetaGen (coding of parameters, fitness function, genetic operators, etc.). In the Genetic Network Design modus (see also Maricic and Nikolov, 1990) in MetaGen, network description files and fitness functions are sent to MetaNet. MetaNet evaluates the networks, returning a fitness value. From MetaNet, standard genetic algorithms can be invoked from MetaGen to optimize a given architecture according to certain criteria. We will not dwell further here on the details of the implementation of genetic algorithms. However, the appendices to this book are devoted to the implementation of neural networks, with particular emphasis on the advantages of modular architectures and local processing.

PART III
Revisiting Modularity

Chapter 7

Evaluation of CALM

7.1 The status of CALM

The CALM model as introduced in Part I is based on a number of basic principles. Of these modularity is the most important one. The architectural principle is derived from the structure of the neocortical minicolumn. The functioning of a CALM module is based on a noise-driven search process structured by competitive selection. This idea finds its basis in the psychological concepts of self-induced arousal (Hebb, 1955) and activation–elaboration learning (Graf and Mandler, 1984; Mandler, 1986). Processing in CALM is further constrained by the demand that it involves only locally available information. General principles such as modularity, locality and noise-driven processes extend far outside the realm of connectionist modelling, as was argued, among others, when discussing genetic algorithms (in Chapter 6). One could even argue that they play a principal role in all self-organizing processes in nature, thus underlying both the emergence of life and cognition. The 'Categorizing And Learning Module' itself must be viewed as a particularization of these more general ideas. The aim of the model as presented in this work is to demonstrate that these principles may be fruitfully combined in connectionist modelling of certain psychological phenomena, and that they may also lead to practical applications. The application of this particular implementation of the general principles was described in Part II.

Based on the above ideas, quite different models could be developed. In fact, in this work, in addition to CALM itself, three different variants of CALM are described. In Part I a variant of CALM, called CALSOM, is discussed that has the ability to organize its representations topologically. In the appendices a discrete version of CALM is reported that uses binary activations, limited time coding of activations, and floating thresholds. Below, we will outline yet another variant that does not employ the strict pairing of V-nodes and R-nodes. Instead, it has fewer V-nodes than R-nodes, which is more biologically plausible. These variant models indicate that many different aspects of CALM can be changed without compromising any of the general principles.

In many places in this work we have mentioned approaches to connectionist modelling that comply with some, but not all, of the general principles. Some of these models have characteristics in common with CALM. Very recently, a study was published by Alexandre *et al.* (1991). They describe a model that is also strongly inspired by the neocortical column. They propose that their model of the cortical column be used as a unit to build large multi-layered networks: 'We thus describe a new connectionist approach, with a processing unit which does not correspond to a neuron but to a cortical column (Burnod, 1988).' (Alexandre *et al.*, 1991, p. 16). It seems that this model adheres at least to two of the basic principles, namely modularity and locality. It is also based on evidence from biology and, to a lesser extent, psychology (the authors prefer to speak of 'information processing'). The simulation studies reported indicate that the model performs well with very fast learning and recognition processes. Another recent biologically inspired model, not mentioned earlier, is by Ambros-Ingerson *et al.* (1990). They simulated layers I and II of the olfactory palaeocortex. Of particular interest is that the model they describe is able to perform hierarchical clustering. Successive 'sniffs' narrow down the representation set for a particular input pattern. The sequence of inclusive sets can be viewed as a clustering structure of the input patterns. We see, thus, that similar biologically inspired approaches have similar characteristics, such as fast learning and the ability to cluster inputs (see Section 3.3).

The relation between neurophysiology and psychology will remain a major research subject for many years to come. CALM must in the first place be viewed as a psychological model. And we fully agree with Mewhort (1990; and so with Hebb, 1958) that behavioural accuracy must be the main arbiter of any psychological model. But we should not stop at that point; exhibiting psychologically plausible behaviour is not sufficient. As Plunkett and Sinha (1991) argue, it is the use of highly structured architectures in psychological modelling that distinguishes (or rather, should distinguish) connectionist modelling from a 'retreat to simple associationism or behaviourism'. In a broad analogy with linguistics, we may speak of 'weak' and 'strong' generative capacity of a connectionist model (e.g. Partee *et al.*, 1990). A model can then be said to be 'weak' if it exhibits (or generates) psychologically valid behaviour, but lacks an internal structure, or if it is impossible to identify the structural characteristics responsible for its behaviour. In a 'strong' model, architectural principles can be identified, they can be related to its behaviour, and they can be accounted for by psychological, biological or computational theory.

In this work we have focused on 'strong' models of behaviour. In these models, the internal structure is not irrelevant. In the remainder of this chapter we will evaluate the plausibility of CALM and of the principles that underlie it. Before discussing some aspects of the psychological plausibility of CALM, we will first address in the next two sections the issues of biological and computational plausibility.

7.2 Biological plausibility

7.2.1 Loosening the R–V-node pairing

Although it is argued that CALM is more biologically plausible than many currently used models, such as backpropagation, some of its features may be criticized from a biological perspective. For instance, there exists no compelling biological evidence for the strict pairing between R-nodes and V-nodes found in CALM modules (F. Crick, personal communication). Earlier in this book we hinted at a functional resemblance between R-nodes and pyramidal cells, and other correspondences between CALM modules and the minicolumns in the neocortex. In fact, the pyramidal cells form about 60% of the cells in a minicolumn (Szentágothai, 1975), which would contradict a strict pairing between pyramidal and inhibitory cells. We do not want to imply that a CALM module describes the functional architecture of a minicolumn in the neocortex. Rather, we want to show that, based on some well-established neurophysiological facts, such as long-range excitation, short-range inhibition, and Dale's law, it is possible to construct a model that has a number of interesting and plausible characteristics.

We will now show how the structure of a CALM module can be modified to bring it into closer agreement with biological reality. We will discuss a class of variants of CALM in which the number of V-nodes is only a fraction of the number of R-nodes. In these modules a single R-node is paired with several V-nodes. The competitive behaviour of the module is preserved, as well as the working of the arousal system. Of course, when each R-node excites several V-nodes (i.e. several up connections, see Figure 2.3), it does not make sense to have mutual, lateral inhibition of the V-nodes. We must remove all such connections (i.e. all flat weights are set to zero, see Figure 2.3). If this is done in a standard CALM module it still works, because the recurrent lateral inhibition (i.e. cross weights, see Figure 2.3) causes *indirect competition* between the R-nodes. Non-zero flat weights merely speed up the convergence process.

Suppose that each R-node excites two V-nodes, and that each such node receives strong inhibition (cross weights) from all other V-nodes. If two R-nodes A and B excite exactly the same two V-nodes, this would present a problem, because then A has no way to inhibit indirectly the activation of B (or vice versa). For one node A to be able to inhibit indirectly another node B in this scheme, there must be at least one V-node that is excited by A and not by B, and that inhibits B. There is full indirect competition if and only if each R-node indirectly inhibits every other R-node. In Figure 7.1, for example, R-nodes A and B are not in indirect competition, because they excite the same two V-nodes. R-nodes A and C, however, indirectly compete with each other. Note that the principle of indirect competition does not imply that all R-nodes inhibit each other *in the same degree*. In general this will not be true if the

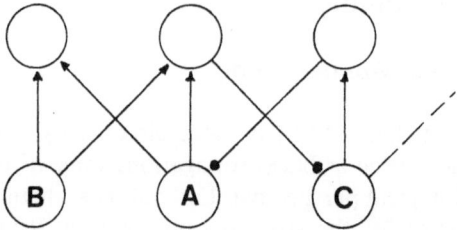

Figure 7.1 Example of indirect competition through recurrent lateral inhibition. Nodes A and C are in indirect competition, but A and B are not.

number of V-nodes is smaller than the number of R-nodes, as we will see below.

We can now describe the construction of a CALM module with indirect competition. Let each R-node excite exactly k V-nodes. Assign a different permutation of k V-nodes to each R-node. All up weights are equal valued, as are all cross weights. Because each R-node is assigned a different permutation of k V-nodes, for any two R-nodes A and B there will be at least one V-node assigned to A but not to B (and vice versa). Through this V-node, A can indirectly inhibit B (and vice versa). Hence, all nodes indirectly inhibit each other, so that we have indirect competition.

Using the above construction scheme, if we have v V-nodes, a maximum of $\binom{v}{k}$ R-nodes can be included in the CALM module. This gives us the possibility of having CALM modules where the number of V-nodes is (much) less than the number of R-nodes. In Figure 7.2, for example, a small CALM module is shown (only the up connections) with six R-nodes and four V-nodes. The wiring pattern is also given in table form. Simulations with this variant CALM module indicate that it takes longer to converge because the V-node activations, which no longer inhibit each other, take rather a long time to decay to such a level that the R-nodes can become activated. The observed

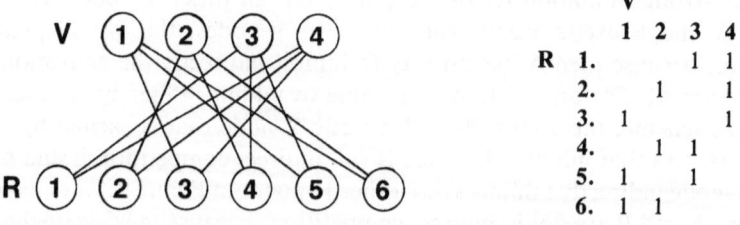

		V			
		1	**2**	**3**	**4**
R	1.		1	1	
	2.	1		1	
	3.	1			1
	4.		1	1	
	5.	1		1	
	6.	1	1		

Figure 7.2 A CALM module with four V-nodes and six R-nodes. Only the up connections are shown (cf. Figure 2.3). The wiring diagram is given to the right. A 1 entry at position (X, Y) means that R-node X excites V-node Y. If an R-node does not excite a V-node, it receives strong inhibition from this node (these lines have not been drawn).

behaviour during convergence is a repetition of short bursts of R-node activity, followed by longer periods of slow V-node decay.

This version of CALM has not yet been tested as thoroughly as the standard module (see Part I), but simulations with the module of Figure 7.2 indicate that its categorization behaviour is largely comparable. When presented with patterns (0 1), (1 0) and (1 1), for example, it correctly classifies them at the first presentation. Categorization of the very first pattern presented takes between 130 and 150 iterations (this involves symmetry breaking). After having learned each pattern for 100 more iterations after convergence, on the second presentation of the patterns, the convergence time decreases to about 65 iterations for (0 1) and (1 0) and to 95 for (1 1) (the latter pattern has to compete with both other patterns). These figures are found using exactly the parameters given in Appendix A (but without down and flat weights). With other parameters, better tuned to the variant module, lower convergence times are found on this task. With up-weight values of 0.50 and an activation decay of 0.25, for example, we found between 90 and 110 iterations for the first convergence and between 13 and 20 iterations at the second (and further) presentations of the pattern (1 0). The module of Figure 7.2 is also able to discriminate between six mutually overlapping patterns.

The arousal system in the module remains operative, because during the competitive phase it still gets brief impulses from the R-nodes. These impulses lessen with repetition of a pattern, because there is less competition. Although during the latent period the V-nodes strongly inhibit the A-node, they do not immediately affect the E-node, so that it takes a while before the E-node has fully decayed. Once a module has converged, several V-nodes remain activated against only a single R-node. This ensures that after convergence the E-node activation asymptotically reaches its zero activation level.

As remarked above, the principle of indirect competition does not guarantee that all R-nodes compete with each other in the same degree. On the contrary, we might say that the price one pays for removing V-nodes is that competition takes on structured patterns. This is nicely illustrated in the simulations. If, for example, the pattern (0 1) is categorized on R-node 3 (in Figure 7.2), then its subsequently presented orthogonal variant (1 0) will most certainly be categorized on R-node 4. R-nodes 3 and 4 are in maximal indirect competition. This can be seen in Figure 7.2. R-node 3 excites V-nodes 1 and 4; R-node 4 receives strong inhibition from these V-nodes. Because the patterns are orthogonal, the second pattern will only weakly activate the V-nodes assigned to R-node 3. In this case, this implies that the remaining two V-nodes have a greater chance of winning the competition, so that the second pattern is categorized by R-node 4.

We could make use of the structured competition by not using all possible permutations (i.e. have fewer R-nodes than possible) and by assigning mainly V-nodes to an R-node 'in the neighbourhood' (see the discussion in Section 3.5). Many such schemes are possible. Suppose, for example, that we have 10 V-nodes, $v(1)$, $v(2)$, ..., $v(10)$, and 30 R-nodes, indexed in groups of three:

$r_1(1)$, $r_2(1)$, $r_3(1)$, $r_1(2)$, $r_2(2)$, $r_3(2)$, ..., $r_1(10)$, $r_2(10)$, $r_3(10)$. Suppose, furthermore, that each R-node excites three V-nodes in the neighbourhood, as follows:

$$r_1(i) \rightarrow v(i-2) \quad r_2(i) \rightarrow v(i-2) \quad r_3(i) \rightarrow v(i-2)$$
$$r_1(i) \rightarrow v(i-1) \quad r_2(i) \rightarrow v(i) \quad r_3(i) \rightarrow v(i+1)$$
$$r_1(i) \rightarrow v(i+2) \quad r_2(i) \rightarrow v(i+2) \quad r_3(i) \rightarrow v(i+2)$$

where '$r \rightarrow v$' means 'r excites v'. It can easily be verified that no two R-nodes excite exactly the same V-nodes. It will also become clear that two R-nodes in a group of three share two V-nodes, and that R-nodes between near groups will usually share less than two V-nodes. In other words, this particular selection of permutations (30 out of a possible 120) induces a sublocal ordering (in the sense of Section 3.5) on the R-nodes, so that we may expect it to exhibit some form of topological organization.

On the basis of such orderings, explanations may be devised for the emergence of topological maps, and for the emergence of competition in general. Such explanations may be more biologically plausible than, for instance, that of Section 3.5, where it was simply assumed that V-nodes have increased inhibition at long ranges. This assumption seems to be somewhat forced. It is therefore fortunate that, in the process of removing the biological implausibility of strict R–V-node pairing, another possible implausibility vanishes as well. A next step would be to study the behaviour of indirect competition with a limited-range, random connection scheme: k V-nodes in the neighbourhood of an R-node are randomly assigned. If the process is purely random there is a slight chance that any two R-nodes will excite exactly the same V-nodes. One might speculate that in biological reality V-node assignment (i.e. basket cells or similar candidates for V-nodes, see for example Crick and Asanuma (1986)) is governed by some process, whereby during early development of the wiring of minicolumns V-nodes are somehow marked, once a synaptic connection has been formed. R-nodes from the same neighbourhood would leave similar marks (possibly caused by the gradient of some chemical over the length of a column). If the pseudopodia of an R-node (i.e. pyramidal cell) tended to avoid V-nodes marked by other R-nodes 'from the neighbourhood', such a growth process would induce the formation of differentiated random R–V-connections of limited range. Because of the marker-guided competitive growth, the V-node connections of the R-nodes would have a strong tendency to be different. In this way, CALM-like structures could evolve with fewer V-nodes than R-nodes. Because of this limitation they would have a tendency to self-organize their representations in topological patterns.

7.2.2 Combining E-nodes to simulate subcortical structures

A further extension of CALM that might increase its biological plausibility would be the combination of all E-nodes of the various CALM modules of a multi-modular network into a single structure. Global excitation of this structure could result from external arousing factors and stimulate (elaboration-type) learning in the entire system. Lesioning this structure would remove the possibility of elaboration-type learning. The simulations in Section 2.4.5, in fact, describe such a simulation, in which the effect of such lesioning was studied in a model of human explicit and implicit memory (for more extensive reports see: Phaf (1991) and Phaf *et al.*, 1990a). It was found that lesioning the E-nodes, which effectively implies removing elaboration learning, results in a strong decline of performance on the simulated explicit memory task (free recall), but not on the implicit memory task (word completion), as indeed is found in anterograde amnesia patients (e.g. Warrington and Weiskrantz, 1970). These results are in agreement with the hypothesis that explicit recall is based on elaboration-type learning, whereas implicit recollection remains possible with only an activation-type learning. Removing elaboration learning would thus be expected to result in a strongly impaired performance on explicit memory tasks, but not on implicit memory tasks. There is considerable evidence that associates anterograde amnesia with damage to the hippocampus (e.g. Mishkin, 1978; Murray and Mishkin, 1984). It would, therefore, seem reasonable to compare the behaviour of the combined E-nodes with certain functional properties of the hippocampus.

7.3 Computational plausibility

7.3.1 Decorrelation by 'negative filtering'

Two of the most basic aspects of CALM, unsupervised learning and modularity, can be defended on the grounds of computational plausibility. Barlow (1989) has argued for a general approach to unsupervised learning in which the redundancy of sensory messages is utilized in the formation of 'representation maps'. The principal goal of these maps is to extract redundancy by filtering out all correlations among the sensory inputs: 'Decorrelating networks based on such principles would "expect" events that occurred in often-repeated sequences, and would tend to respond more strongly to abnormal ones.' (Barlow, 1989, p. 300). A map may be regarded as a 'negative filter' (Barlow, 1989), through which incoming messages are automatically passed. The expected elements are 'subtracted' from the incoming messages, and only the novel or unexpected is let through. For this to be possible it is crucial that the system acquires the *a priori* probabilities of its input patterns. In other words, the system must learn to distinguish between old and new patterns. The combination of categorization and unsupervised learning in CALM fulfils these objectives. Assigning patterns to categories based on relative pattern distances is an effective way to decorrelate patterns. In a CALM module, the similarity between any patterns A and B, based on the co-occurrence of a sufficient number of elements in both patterns, may cause A and B to be categorized under a single representation. Strongly correlated patterns are merged, whereas sufficiently different patterns are separated. The dynamics of this categorization process are such that prior learning determines whether an input pattern is considered sufficiently novel and unexpected and should be assigned to an independent category, or whether it should instead be considered a version of some earlier presented pattern. Thus, the learning of input activation patterns causes a CALM module to set up a 'negative filter', which becomes embodied in the modified weights.

Learning *per se* is no guarantee that the modified weights will optimally reflect the prior probabilities. Prolonged or repetitive exposure to certain patterns may cause the prior probabilities of these patterns to reach their equilibrium values quickly. Frequency should have an effect on learning, but this effect has to be constrained. It is not sufficient that a constant change in the representation map takes place at every presentation, because this would drive the weights to their equilibrium values in too short a time. The module would then react strongly to each individual pattern and it would lose much of its capacity to generalize over similar patterns by clustering them under a single categorization. In CALM, weights are adjusted with a rate that depends on the (relative or effective) novelty. This property of CALM can be related to the more general property of habituation found in almost all biological systems. Habituation mechanisms can be argued to yield adaptive changes that reflect the inverse frequency of the event that caused the adaptive behaviour. In CALM, the E-node activation, which controls the learning rate, habituates because it reflects the novelty of input patterns. Thus, the categorization and learning mechanism in CALM causes input patterns to become decorrelated in a 'negative filter', that is maintained through novelty-dependent learning.

7.3.2 Modularity and storage capacity

Some authors have criticized the use of a winner-take-all mechanism in CALM and other modular structures (e.g. see Vogl, 1989; but see also Murre *et al.*, 1989b). One of the arguments that appears to weigh against the use of localized architectures is that it may seriously limit the storage capacity. We agree that in principle this is a valid argument *if* the only objective is to store as many patterns as possible in a given number of nodes. But this can never be the only goal of a biologically plausible system. As we have argued above, the combination of categorization and learning in a modular structure may cause input patterns to become decorrelated, and provide the system with a sense of what to expect, a way to differentiate between the significant and the trivial. In order to enforce such a categorization, a selection mechanism must be implemented. Modules with a winner-take-all competition such as CALM are very suitable for this purpose. The approach taken in CALM is not to be viewed as purely localized, however, because we envisage CALM models to consist of many modules. When the modules are considered as elements of a larger system, the representations of input patterns are in fact distributed over as many nodes as the number of modules involved in processing the pattern. Modular architectures in this view stand somewhere in between purely localized and fully distributed representations. They allow considerable control over the massively parallel processing, while retaining a sufficient memory capacity to be used in realistic tasks. If, for example, one layer of several modules is considered, the number of different outputs from such a layer (i.e. the number of discriminable patterns) is not simply the sum of all nodes in the modules. All possible combinations of one node per module should be counted. This means that the potential storage capacity of m modules with n nodes equals n^m. Twenty modules, each of size 10, would in principle suffice to store 10^{20} patterns, which incidentally equals the total intake of information in a lifespan as estimated by Von Neumann (1958). It appears then that even suboptimal utilization of the storage capacity of neural networks ensures a more than sufficient memory capacity.

An even stronger argument against the requirement of maximum storage capacity is given below, where it is argued that any constraints on the architecture (i.e. those reducing storage capacity) will improve the set generalization of a network. Moreover, the learned behaviour of constrained networks will more likely persist when confronted with novel inputs. We may thus conclude that both from a biological and a computational point of view the maximum storage counter-argument cannot be defended.

7.3.3 Extendibility and independent functioning of modules

An important characteristic of CALM networks concerns their extendibility. A CALM module's functioning is relatively independent of the rest of the network in which it is embedded. So, adding a module to a network will not fundamentally change the network's behaviour, whereas adding a comparable number of nodes to a homogeneous network will have a considerable effect on the functioning of the network and may call for extensive retraining of the entire network in order to restore (and perhaps improve) the already learned behaviour (e.g. Waibel, 1989). In principle, the modular design enables the unlimited extendibility of a network, when sufficient resources (i.e. computer memory, etc.) are available. The brain appears to use the same design principle with a large number of small columns. Columns appear to consist of no more than about 100 neurons. A comparable size also seems plausible for a CALM module. Though categorization speed only decreases slowly for larger modules, there will be a finite probability that no suitable representation can be found with very large modules (see Chapter 3). We may thus speculate that the brain avoids possible convergence problems by keeping the columns relatively small. It should be noted, however, that the occasional absence of a convergence in some modules does not constitute a fundamental problem in a modular network, because processing in other parallel streams may still lead to convergence at higher levels. This is in fact the mechanism used in the inter-active activation model by McClelland and Rumelhart (1981; see also Section 4.1). In homogeneous networks the inability to process part of the information would delay or even halt further recognition of the pattern presented.

7.3.4 Limiting structure to improve the quality of solution

Another argument in favour of limiting the structure of a network derives from considerations of retrieval efficiency, or quality of solution. Recent evidence suggests that for models with full connectivity between nodes, such as Hopfield networks or the Boltzmann machine, the quality of the solution may be improved by reducing the connection-to-node ratio. Barna and Kaski (1990) have shown that with the Boltzmann machine much better solutions were achieved on the T-C problem and the encoder problem after randomly removing a fraction of the connections (both problems are described in Rumelhart *et al.* (1986); for the encoder problem, see also below). In the case of the T-C problem on a 7×7 torus, they obtained the best results if 75% of the connections were randomly deleted (before learning). The demonstration that even random limits on the connectivity may improve performance further reinforces our conviction that constraining the architecture of networks may result in faster and more efficient models.

Another study that shows how limiting structure may improve the quality of solution is reported by Rueckl *et al.* (1989). They used a three-layer back-propagation network that received 25 inputs from a 5×5 input grid. On this grid nine square 3×3 patterns could be written at nine possible locations. The output consisted of 18 nodes, nine for form identification and nine for position identification. The model had 18 hidden nodes. When only four to six of the hidden nodes were connected to the nine position output nodes, and the remaining nodes to the form identification nodes, the model attained a significantly better performance than when full connections between the hidden layer and the output layer were used. The authors relate this finding to the early separation of position and form information as can be found in the cortical visual system.

We thus see that limiting the structure of neural networks may improve several aspects of their behaviour. In the next section, we will demonstrate that, by reducing their representational capacity, the sometimes 'catastrophic' retroactive interference may also be reduced. This problem has led several researchers to discard neural networks as interesting models for human memory processes. Before arguing in the next section why this conclusion may have been drawn too hastily, we will very briefly review another important aspect of neural networks that can be proven to benefit from constrained – and most probably also from modular – architectures, namely set generalization.

7.3.5 Network architecture and generalization

The term 'generalization' may give rise to confusion, because in psychology and computational learning theory it refers to different concepts. In this work for example, in Part I we have used it in a psychological sense to indicate the case where several patterns are represented by a single Representation-node. A CALM module is forced to generalize, in this sense, when it is presented with more distinct patterns than it has R-nodes. As was shown in Section 3.3, in such cases similar patterns will be grouped together in the same category. In other words, the CALM module is capable of performing clustering of the input patterns. In doing so, it must generalize over patterns.

The concept of generalization stems from a long tradition in psychology studying the likelihood that a response to a given stimulus A will generalize to another stimulus B. Shepard (1957, 1987) has convincingly argued that such generalization behaviour is subject to a very general law that applies to many, if not all, 'sentient organisms' and to most stimulus domains. An important precondition for application of Shepard's universal law of generalization is that the stimulus domain first be transformed to a psychological space. According to the law, the probability of generalization decays exponentially with distance in this psychological space. In response to Shepard's seminal work, a large body of research has emerged focusing on generalization in experiments on learning, identification and categorization. For example, Nosofsky (1985, 1986; see also Nosofsky et al., 1989) has studied how subjects may 'stretch' psychological space to give more prominence to important stimulus dimensions.

Gluck and Bower (e.g. Gluck, 1991; Gluck and Bower, 1988) have recently developed connectionist models that exhibit the generalization behaviour described by Shepard's law. In their models, they use the Widrow–Hoff delta-rule (1960), which as they point out is formally equivalent to the model of classical conditioning proposed by Rescorla and Wagner (1972). Although the model by Gluck and Bower is very interesting and exhibits the desired behaviour, the authors cannot account for most architectural aspects of their model in terms of their psychological function or relation to biological structures. As we have argued at the beginning of this chapter, these aspects are essential for 'strong' models of behaviour. Also, Kruschke (1990) has developed a connectionist model of categorization based on Nosofsky's theories. The same remarks apply to Kruschke's model.

Here, we want to focus on generalization in a more computational framework. With *set generalization* we mean the likelihood that a neural network that is trained to criterion on a certain pattern set will continue to exhibit this behaviour. The question we consider is: 'Will the learned behaviour persist and generalize to new patterns?' We are particularly interested in the relation between network architecture and generalization behaviour. Solla (1989), for example, has demonstrated that reducing the receptive fields of hidden nodes on a certain task may greatly improve the generalization behaviour. This study forms another case where constraining the architecture improves its performance. Solla (1989), with reference to Denker *et al.* (1987), explains this as a consequence of a reduction of the internal entropy of the network. Roughly speaking this means that the prior probability of producing the desired mapping of inputs to outputs (i.e. through learning) is reduced, in this case by constraining the architecture of the network.

Wolpert (1990b), who approaches the question of generalization in a very general manner, subsumes architectural characteristics such as 'internal entropy' under the heading 'measures of simplicity' (e.g. his Table 1). This measure of simplicity is related to Occam's Razor, the principle of parsimony that says that entities are not to be multiplied without necessity, or in the words of the philosopher: 'It is vain to do with more what can be done with fewer'. For the case of neural networks, it implies that if two networks have learned a given pattern set to criterion, we may expect the 'simpler' of the two to do better in the future. In the past few years, several studies have appeared that propose to use the Vapnik–Chervonenkis dimension (VC-dimension) as a basis to assess the 'simplicity' of a neural network and to predict its generalization behaviour. It appears that this particular measure of simplicity gives better results than other measures, such as internal entropy.

The VC-dimension refers to the 'expressiveness' of a network. In the learning theoretical approach, a neural network is seen as a combination of an architecture and learning method, together forming a 'system for hypothesis formation'. Imagine the task where a network is presented with 1000 X-rays. Some of the X-rays contain malicious tumours. For each X-ray the network is instructed whether a tumour of a specific type is present. After sufficient training it is able to classify correctly 950 X-rays (i.e. correctly saying whether a tumour is present or not). To test it, the trained network is then presented with 100 novel X-rays. What is the probability that it will classify correctly, say, 90 of these X-rays? To estimate this probability we must know the VC-dimension of the learning system (or some other 'measure of simplicity'). If, for example, the system were a standard computer memory with sufficient storage capacity to remember exactly 950 X-rays, we would expect virtually no generalization behaviour. This system would have a high VC-dimension. If, on the other hand, we knew that a very small neural network were used, we would probably have been surprised that it was able to learn the 950 X-rays in the first place, and we would expect this behaviour to persist and to generalize to other, similar patterns. The small neural network has a low VC-dimension.

The general format of the above learning task has become known as probably approximately correct learning or PAC learning (Haussler, 1990) and was first proposed by Vailliant (1984). A learning method is not restricted to the domain of neural networks; the above framework is appropriate for studying many different algorithms for machine learning (see Haussler (1990) for a review). Above, we were interested in the probability that on future occasions the network would still correctly classify a certain fraction of X-rays. The effect of learning (i.e. the formation of sensible hypotheses) in PAC learning is usually formulated in terms of the probability $1 - \delta$ that the method will correctly classify at least a fraction of $1 - \varepsilon$ of the testing patterns (Blumer *et al.*, 1989). In the X-ray example $1 - \varepsilon$ would be 90 out of 100, so ε would be 0.10, whereas $1 - \delta$ is the probability we want to know.

The VC-dimension is the smallest number d at which a learning system starts failing to induce all possible 2^N binary functions on any N examples (e.g. Abu-Mostafa, 1989; Blumer *et al.*, 1989; with reference to Vapnik and Chervonenkis, 1971). Suppose, for example, that we have a learning system for concepts or word meanings, consisting of m atomic representations, such as [+living], [+human], [+animal], etc. Each word meaning can be represented by a set of such atomic representations. The word 'Dog', for example, can be represented by the concept {[+living], [+animal], [+carnivore], ...}. Call C the set of all concepts that can be formed by the system (i.e. some concepts may be excluded, see below), and call the set of all constituting atomic representations X. Take any subset S of atomic representations from X. If C can represent any concept in S, we say that C 'shatters' S. In our example, all combinations of atomic representations are allowed. Hence, C can represent 2^m concepts, where m is the number of atomic representations in X. We say that, in this case, C shatters X, and the VC-dimension of C is equal to m.

Concept representation systems often have internal constraints excluding, for example, combinations such as [+human] and [+elephant]. Such a system cannot shatter any subset S from X. It could not, for example, shatter a set S that includes both the atoms [+human] and [+elephant], because the set of concepts does not include any concept with this combination. In these cases, to determine the VC-dimension of the learning system, we must find the largest set S in X that is still shattered by the system (i.e. that does not exclude any combinations in S). The size of this set is the VC-dimension (Blumer *et al.*, 1989).

So, we see that constraining the representational capacities of the learning system reduces the VC-dimension. Within the framework of PAC learning, using the VC-dimension as a measure of simplicity, Baum and Haussler (1989) have derived several measures of generalization behaviour for neural networks. The question they are trying to answer is as follows. How many example patterns m should one present to a network so that at least a fraction $1 - \varepsilon/2$, for $\varepsilon < 1/8$, is correctly classified at training, and so that one may expect with high probability the network to classify at least a fraction $1 - \varepsilon$ patterns on testing? For a feedforward network of binary nodes (with a linear threshold activation function, i.e. using the Widrow–Hoff (1960) learning rule) with a single output node they have derived the following value. Under the above conditions, if a network with N nodes and with W weights (including learning thresholds) is able to learn m example patterns, where

$$m \geqslant \frac{32W}{\varepsilon} \ln \frac{32N}{\varepsilon} \qquad (7.1)$$

one has a confidence of $1 - 8e^{-1.5\,W}$ that at least $1 - \varepsilon$ will correctly be classified (Baum and Haussler, 1989, Corollary 4, p. 155).

Going back to our X-ray example, could we now say something about the adequacy of training? Suppose we use a network with binary nodes for training with five nodes and with 50 weights (assuming considerable preprocessing of the X-rays). As we remarked above, in this example ε is equal to 0.10. Under these conditions we derive that m, the minimal number of training examples, must be at least 118 044, which is a very large number. As Baum and Haussler (1989) acknowledge, the bounds given by equation (7.1) are a great overestimate. In addition to this, the approximation concerns a worst-case analysis and makes no assumptions about the distribution of patterns. More detailed analyses are necessary to put these results of computational learning theory to practical use. Ideally, we would like to have some method of assessment that would take as input a formal description of the architecture, some parameters describing the pattern domain (including the number of patterns), the desired learning criterion (i.e. ε), and giving as output probability measures on various levels of generalization.

We can nevertheless draw an important conclusion from the above results, namely that constrained architectures will have better generalization behaviour. This is sometimes even true when the architectures are reduced *after* training, for example by removing near-zero weights and nodes with only near-zero input weights (Baum and Haussler, 1989, Corollary 6). As we have argued in Part I and the appendices, modularity in neural networks will reduce the number of learning connections, so that modular neural networks may be expected to have better generalization characteristics than comparable non-modular networks. The reader should be aware that this assertion starts from the assumption that the network is able to learn the training set. Of course, if the architecture is constrained too much, this may hamper learning behaviour itself, to the point that it can be proven that a certain set cannot be learned by the given architecture. The question of generalization only becomes interesting *after* a network has managed to learn a set of patterns to criterion.

From a generalization point of view, we would like to use the simplest network (e.g. according to equation (7.1)) that is still able to learn a given pattern set. This requirement may sometimes conflict with other demands, such as a large storage capacity. But as we have argued in this section, in many cases, constraining an architecture in a suitable way may improve not only its generalization behaviour, but also other desirable characteristics such as extendibility, retrieval efficiency and learning time. In the next section we will see that it may also reduce sequential interference.

7.4 Psychological plausibility

With a slight variation on Popper (1972), we might say that it is probably impossible to demonstrate that a certain model is biologically, computationally or psychologically plausible, but that it remains possible to show that the model is at least not subject to certain types of implausible behaviour. In Part II, we have shown that for certain psychological experiments in implicit and explicit memory, CALM-based models are psychologically plausible models (see also Phaf, 1991; and Phaf *et al.*, 1990a). In this section we will focus mainly on another aspect of memory: catastrophic sequential interference. This problem plagues many existing neural network models. We will argue that constraining architectures, including the introduction of modularity, may reduce this interference.

It has also been argued from a general perspective that neural networks, as a class of models, can never be plausible models of human behaviour (e.g. Fodor and Pylyshyn, 1988; Levelt, 1989). We will not review this discussion here. One aspect, however, does seem to warrant some attention. Sometimes, an argument in the discussion about the psychological plausibility of connectionist models is heard that asserts that they are merely finite-state automata (e.g. Levelt, 1989, p. 213). It can easily be shown, however, that the generating power of a neural network with continuous activation values, even if they are bounded, is strictly stronger than a finite-state automaton. The argument below is illustrative of this.

7.4.1 Infinite-state neural networks are not finite-state automata

A neural network with binary nodes has only a finite number of states and is, therefore, not able to recognize a context-free language. It lacks the infinite-size stack that enables the push-down automaton to recognize embedded constructs in context-free languages. It is, however, possible to construct a neural network that is able to recognize a context-free language, if it has nodes with an *infinite* number of states. A straightforward proof of existence of this is based on the use of 'step-up step-down' nodes.

A 'step-up step-down' node has an activation value with a (countably) infinite number of states. Its activation value, a_t, may be either bounded or unbounded. Its defining characteristic is that a positive input, $+I$, at time t, followed by a negative input, $-I$, at time $t+1$, results in an activation $a_{t+1} = a_{t-1}$. In other words, at any point in time, an input can be cancelled by an equal input with opposite sign.

Let f be a continuous, increasing function with an inverse f^{-1}, and let the activation rule be given by

$$a_t = f(f^{-1}(a_{t-1}) + I) \qquad (7.2)$$

That is, the activation at time t is dependent on the inverse of the activation at time $t-1$ increased by an input signal I. For example, an unbounded node could be based on the function

$$f(x) = x \qquad (7.3)$$

with

$$a_t = a_{t-1} + I \qquad (7.4)$$

A bounded neuron could be based on the function

$$f(x) = (1 - e^{-x})^{-1} \qquad (7.5)$$

and

$$f^{-1}(x) = -\ln(1 - x^{-1})^{-1} \qquad (7.6)$$

with

$$a_t = 1/[\![\, 1 - \exp\{[1/-\ln(1 - a_{t-1}^{-1})] - I\}\,]\!] \qquad (7.7)$$

A node with an activation rule as given by equation (7.2) has the 'step-up step-down' property.

Proof. Assume that at time t a quantity $+I$ is present as input and at $t+1$ a quantity $-I$. Then we have

$$a_t = f(f^{-1}(a_{t-1}) + I) \qquad (7.8)$$

$$a_{t+1} = f(f^{-1}(a_t) - I) \qquad (7.9)$$

$$= f(f^{-1}(f(f^{-1}(a_{t-1}) + I)) - I) \qquad (7.10)$$

$$= f(f^{-1}(a_{t-1}) + I - I) \qquad (7.11)$$

$$= f(f^{-1}(a_{t-1})) \qquad (7.12)$$

$$= a_{t-1} \qquad (7.13)$$

∎

Consider a 'step-up step-down' node with two incoming weights, w^+ and w^-, with fixed values of opposite sign. This simple neural network recognizes the language $w \in \{a,b\}^*$, $\{w/ \#_a w = \#_b w\}$, i.e. the language consisting of strings with an equal number of symbols a and b. This language is well known to be context free (and not regular). Suppose we present the symbols a over one connection, say with weight w^+, and the symbols b over weight w^-. The symbols are presented subsequently to the node, one symbol per iteration, starting at iteration $t+1$. It will be immediately clear that presenting an a will increase the activation by a certain amount, and that (only) presentation of a b will result in a similar decrease. The recognition criterion, therefore, is that after presentation of all symbols in the string, the activation should be equal to a_t, i.e. the activation at iteration t. Only if this criterion is met is the string recognized. A single 'step-up step-down' node is, thus, able to recognize a context-free language.

It is also possible to recognize the context-sensitive language $\{w/ \#_a w = \#_b w = \#_c w\}$ by combining three of these nodes, and by defining a joint recognition criterion. Connect, for example, input nodes for a and b to node 1, a and c to 2, b and c to 3, and demand that after presentation of a string each of the nodes is back in its initial activation state.

The 'step-up step-down' node does not have a stack, and is therefore a very limited recognizer. Contrary to the push-down automaton, it cannot recognize any given context-free language, because it lacks the ability to remember the position of a symbol, which is a crucial characteristic for a stack. It has, however, one property in common with the push-down automaton: it is able to count. And this suffices for recognizing at least a subset of the context-free language.

Counting to infinity presupposes infinitely precise activations. This requirement is no less implausible than the infinite tape length of the Turing machine. Although we would not argue that this 'neural network' is in any way psychologically plausible, it provides a straightforward illustration that neural networks with infinite-state nodes are not merely finite-state automata. In Section 5.1, several connectionist models were discussed that suggest that it is indeed possible to construct connectionist models that incorporate language or music grammars, and that human learning of these grammars, the syntactic inference problem, may perhaps be modelled by learning neural networks. We are at the moment developing a CALM-based model that is able to recall branching sequences. The architecture is based on Baddeley's articulatory loop (Baddeley, 1986; Baddeley and Hitch, 1974; see also Burgess and Hitch (1991) for a recent study that is similar in spirit to our approach). This model, called ELAN-10, has already been shown to exhibit psychologically plausible recall characteristics (e.g. it shows a primacy effect in serial recall). It is, furthermore, able to learn, say, two sequences *abc...* and *abd...* under two different contexts. If after training, it is presented with the sequence *ab* and a context, the model is able to finish the sequence in accordance with context (i.e. either producing a '*c*' or a '*d*'). The model, thus, is able to do some basic grammatical inference. We intend to extend this behaviour to the point where the model is able to learn to perform simple reasoning (e.g. basic arithmetic). We will report on this work elsewhere. In the remainder of this chapter we will focus our attention on one of the major points of critique on the psychological plausibility of neural networks: catastrophic interference.

7.4.2 Retroactive interference

One of the major weaknesses of many connectionist models is that on sequential learning tasks they exhibit very strong retroactive interference: newly learned patterns may erase nearly all existing memories (Grossberg, 1987b; McCloskey and Cohen, 1989; Ratcliff, 1990). This catastrophic interference has been mainly investigated in simulation studies with backpropagation models, which seem to be especially prone to this psychologically implausible behaviour. Several aspects of a simulation may influence the amount of interference, in particular model architecture (including learning rule or algorithm, model parameters, etc.) and pattern presentation (method of encoding, order of presentation, length of time with which each pattern is presented, etc.).

Other important aspects include the structure of the training and testing patterns (e.g. four-letter words give less interference than random bit strings; Brouse and Smolensky, 1989) and the correlation between the two domains to be associated. With auto-associative learning, for example, there is a perfect correlation of input and output domain, so that inputs are highly predictive of outputs. Hetherington (1990) argues that such domain correlations may contribute towards an increased ability to generalize and a decreased sensitivity to interference. Before discussing the importance of modular and other semi-distributed architectures for reducing interference, let us briefly review the simulations by McCloskey and Cohen (1989), followed by some attempts to remedy the problem by improving the pattern presentation scheme.

Pattern presentation and interference
The study by McCloskey and Cohen (1989) aimed at teaching a model (by 'rote learning') some simple arithmetic: adding, subtracting, dividing and multiplying numbers in the range zero to nine. During training, two numbers and an arithmetic operator were presented as input patterns, while the correct answer was presented as output. The model consisted of a straightforward, three-layer backpropagation model (Rumelhart *et al.*, 1986) with fully connected layers. Numbers (single digits) were represented by activating three consecutive nodes in the output or input layer. The number 3, for example, was represented as 0 0 0 1 1 1 0 0 0 0 ..., and a zero as 1 1 1 0 0 0 0 .... The input layer consisted of 28 input nodes (two times 12 nodes for representing the digits, and four additional nodes for the operators), the hidden layer consisted of 50 nodes, and the output layer consisted of 24 nodes (12 for digits and 12 for tens).

The network could easily be taught all summed digit pairs, as well as all multiplied digit pairs. The pairs were presented for training in blocks with varying random order. When the network was trained on patterns drawn from the entire training set, no problems occurred. But when the network was first taught all additions with one (e.g. [1 + 1], [2 + 1], [3 + 1], ..., and also [1 + 2], [1 + 3], [1 + 4], ...), and only *then* on all additions with two (except [1 + 2] and [2 + 1], which had already been learned), the newly learned patterns appeared to have washed out all memory of addition with one. Performance on the ones dropped from 100% to 57%, after a single run, and to 30% after two runs on the twos.

The simulations by McCloskey and Cohen (1989) were replicated by Hetherington and Seidenberg (1989) with essentially similar results. A second simulation by these authors, however, indicated that learning the twos does not completely destroy all memory of the ones. It appeared that the ones were relearned faster than a totally novel set of additions (with three). The model thus showed evidence of 'savings' (Ebbinghaus, 1885): the ones were not completely unlearned. Based on these results, Hetherington and Seidenberg (1989) argue that the catastrophic interference found by McCloskey and Cohen (1989) is primarily dependent on the method of pattern presentation. In particular, they argue (p. 30) that *blocking* of learning trials (i.e. first a block of ones, then a block of twos) may be an important contributing factor, and that 'if, instead of following this strict blocking scheme, there is some minimal retraining on the ones, performance will rapidly improve due to savings'. Based on this idea they used a training scheme intermediate between both strict blocking and fully concurrent presentation of patterns.

Hetherington and Seidenberg (1989) trained their model in five stages on addition with ones, twos, threes and fours. For each addition, a set was constructed containing 13 digit pairs as mentioned above. The sets were constructed so that they would not overlap (i.e. [1 + 3] occurred in either the set of ones, or the set of threes, but not in both). The training scheme is shown in Table 7.1. Presenting two sets of one type corresponds to presenting each element of the set twice, in random order, possibly interleaved with elements from other sets. From the table it becomes clear that the consecutive stages overlap, so that retraining of patterns does occur. At stage 5, the ones are no longer retrained, so that on the basis of the above-cited data we might expect considerable interference as a result of training the twos, threes, fours and fives. The results, however, indicate that this is not the case. After training on the last stage, the model is still able to reproduce correctly in between 12 and 13 ones (out of a possible 13). The authors further report that after continued training for 35 more epochs following stage 4, the mean number of correct responses on the ones was still 91% (11.8 out of 13). Their conclusion, therefore, is that this method of pattern presentation prevents catastrophic interference.

We did a series of simulations to investigate further the effects of pattern presentation schemes on retroactive interference. Our findings indicate that Hetherington and Seidenberg's (1989) results on the detrimental effects of strict blocking do not directly generalize to other models. A method similar to their 'method of overlapping stages', however, appears to work well on auto-associative learning of random pattern vectors. The patterns used in our simulations consisted of eight elements. Ten new patterns were generated for each simulation (and replication). Each of the pattern elements was assigned a

Table 7.1 Training scheme used by Hetherington and Seidenberg (1989). The table shows the sets presented in each stage. A stage lasted for 10 epochs. In each epoch, patterns were presented in a different random order.

Stage	Pattern sets							
1	1	1						
2	1	1	2					
3		1	2	2	3			
4		1	2	2	3	3	4	
5				2	3	3	4	4	5

(uniform) random value between zero and one. The length of the vector was normalized to 1.0. The model used was a simple three-layer backpropagation network. The size of the input and output layers was eight and the size of the hidden layer was five nodes (simulations indicated that increasing the hidden layer beyond this size did not essentially influence the results). Before every simulation, weights were (uniformly) randomly initialized in the range $[-0.5, 0.5]$. The learning rate was 0.5 and the momentum parameter was set at 0.9. With these parameters, the network easily learns 10 random patterns to the criterion described below.

Simulation 7.1. In this simulation, 'strict sequential learning' was used. Each of the patterns was learned until the criterion was reached, and *not repeated thereafter*. The criterion consisted of a correlation coefficient (i.e. the cosine of the angle between the two vectors) of more than 0.99 between the (target) pattern and the pattern produced at the output layer. The simulation was repeated 100 times. Each time both the initial weights and the patterns were generated anew. The averaged results are shown in Figure 7.3. Recall is represented by the correlations remaining after having learned all patterns. The base rate shown in the figure is the expected correlation of 0.863 between a random pattern and its output before the network has learned anything. It was established by generating 5000 random patterns and averaging the correlations. As

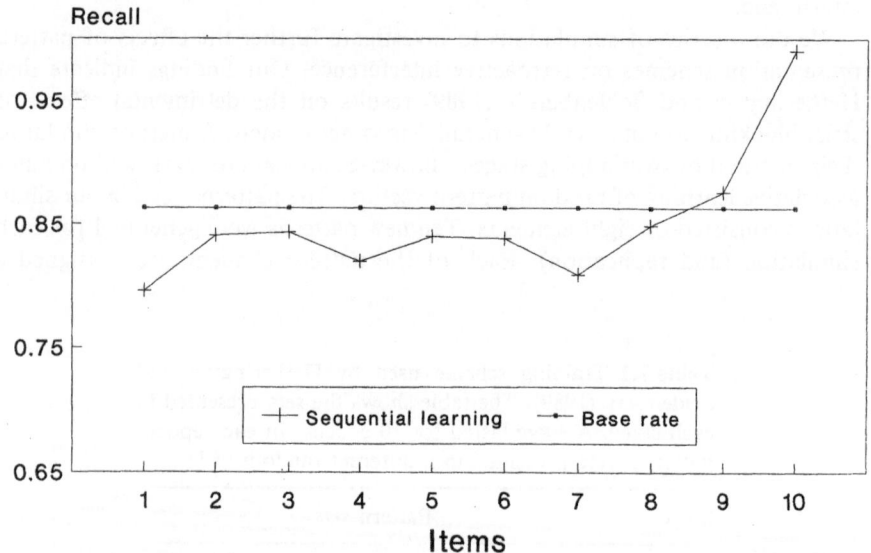

Figure 7.3 Interference in backpropagation as a result of strict sequential learning. The items are normalized random vectors. Recall was found by measuring the output correlations after learning. Base rate is the correlation to be expected before any learning has taken place. The results are averaged over 100 replications.

can be seen from Figure 7.3, strict sequential learning causes catastrophic interference to the extent that after learning the network actually performs *worse* than before learning.

Simulation 7.2. Having established that in this simulation paradigm strict sequential learning gives rise to 'more than catastrophic interference', we trained the network using 'strict blocking' of trials. First, five patterns were simultaneously trained until the criterion was reached (see above), followed by training on patterns 6 to 10. After these had reached the criterion, the network was tested for recall. The simulation was repeated 100 times. The results are shown in Figure 7.4. Strict blocking also leads to considerable retroactive interference, although not as bad as strict sequential learning.

Simulation 7.3. To test whether training in 'overlapping stages', as described by Hetherington and Seidenberg (1989), is a feasible method for reducing interference the following training method was used. A 'window' was moved over the ordered pattern set. All patterns in the window were trained to the set criterion. Let us say that the window can contain three patterns (we will speak of a *depth* of 3). Then the training stages are as follows. Stage 1: pattern 1; stage 2: patterns 1 and 2; stage 3; patterns 1, 2 and 3; stage 4: patterns 2, 3 and 4; stage 5: patterns 3, 4 and 5; etc. Simulations were carried out with windows of depth 1, 2, 3 and 4. Network, patterns and parameters were as above. For each depth, 100 simulations were carried out. The averaged results

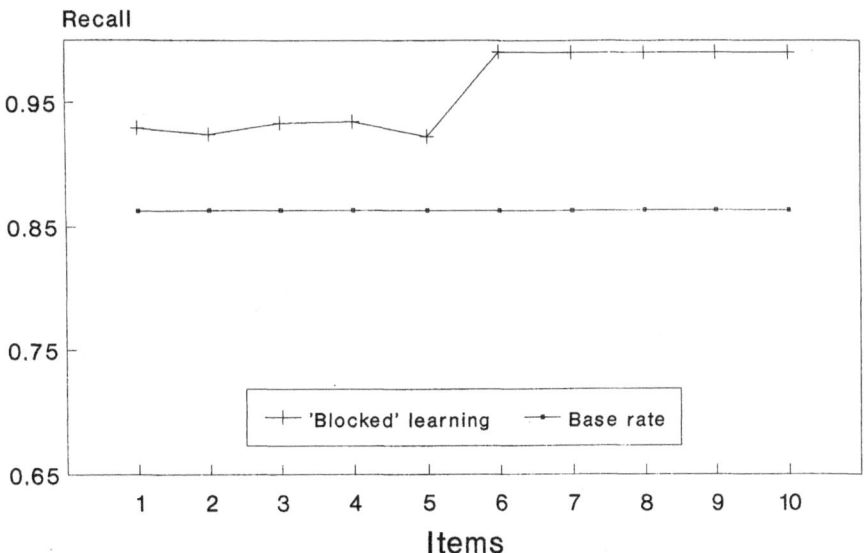

Figure 7.4 Interference in backpropagation as a result of a strict blocking scheme. For an explanation see Figure 7.3.

are shown in Figure 7.5. A depth of 1 leads to strong retroactive interference, comparable with using zero depth (i.e. strict sequential learning). Depths of 2, 3 and 4, however, lead to a progressively decreasing interference, although even a depth of 4 performs hardly better than strictly blocked learning in this respect.

Simulation 7.4. According to Hetherington and Seidenberg (1989), the overlapping stages method leads to reduced interference because old patterns are occasionally retrained. The 'windowing method' of the previous simulations only rehearses the most recent patterns. It would be interesting to see whether an increase in performance (i.e. a reduction in interference) could be achieved by rehearsing a constant number of patterns chosen randomly from the already learned patterns. In the method used, the first patterns have a higher chance of being rehearsed than late patterns in the list. The exact chances of rehearsal with list size 10 are shown in Figure 7.6. With a depth of 2, for example, pattern 3 will on average be rehearsed about four times (out of a possible 10). The term *depth* is maintained, although here it refers to randomly selected items. We remark, furthermore, that if the depth is 3, it implies that the first four items will certainly be rehearsed up until pattern 4. The results are given in Figure 7.7. As can be seen, the 'random rehearsal method' works successfully. Especially with depths of 3 and 4 retroactive interference is strongly reduced.

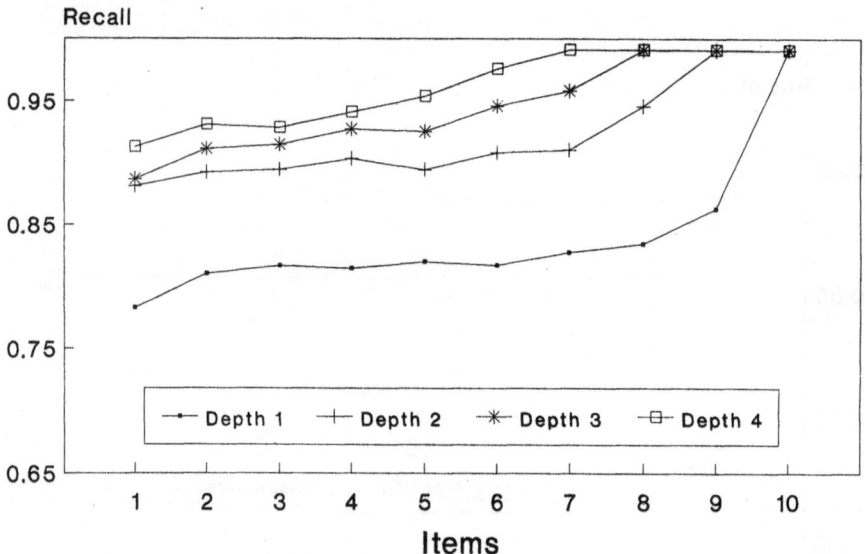

Figure 7.5 Interference in backpropagation using a windowed training method. Depth refers to the number of most recent patterns being presented simultaneously with the current pattern. Recall was as in Figure 7.3. Each point is averaged over 100 replications.

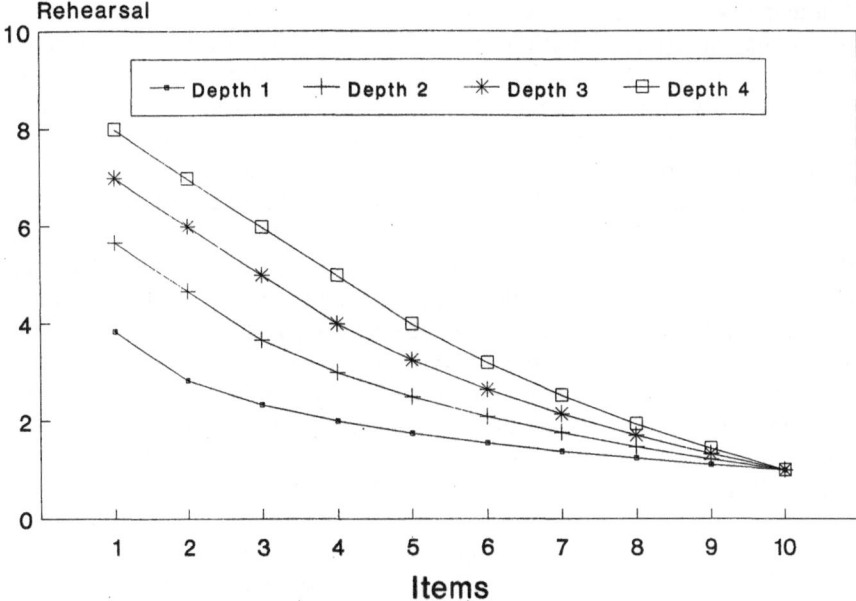

Figure 7.6 Expected number of item rehearsals (out of a possible 10) for depths of 1 to 4. See text for an explanation.

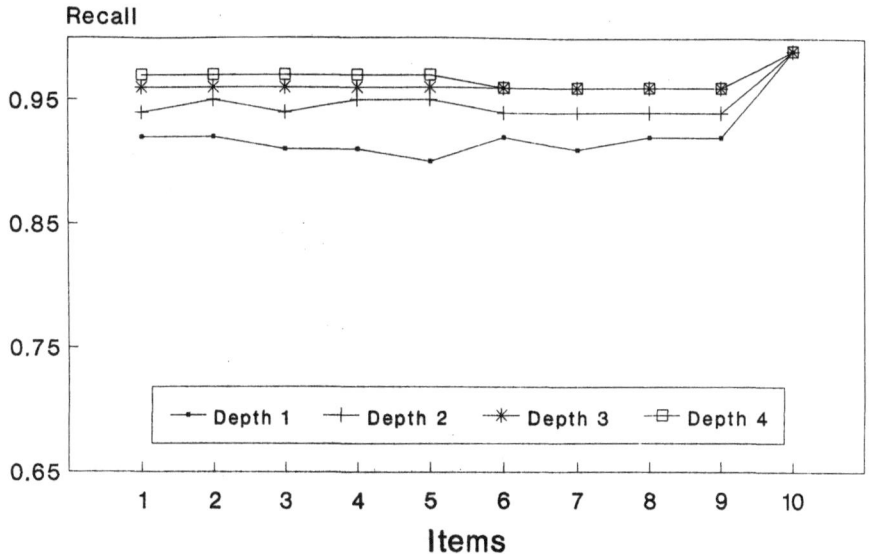

Figure 7.7 Interference in backpropagation using a random rehearsal method. Depth refers to the number of previous patterns randomly selected for rehearsal, simultaneously with the current pattern. Recall was as in Figure 7.3. Each point is averaged over 100 replications.

Simulations 7.1–7.4, and the other studies cited above, convincingly demonstrate that pattern presentation schemes may considerably influence retroactive interference, from 'more than catastrophic' for strictly sequential learning to 'only slightly' for the random rehearsal method. What is surprising in these studies is that important architectural characteristics such as the size of the hidden layer have a negligible influence on retroactive interference. In Figure 7.8, for example, a replication of Simulation 7.1 is shown with hidden layers of 5 and 20 nodes. As can be seen, increasing the size of the hidden layer only has a minor influence on retroactive interference (this was also found by Hetherington and Seidenberg (1989); see their note 3, p. 32). In fact, increasing the hidden layer seems to have a slight *negative* effect (see below for a possible explanation). The negligible influence of the size of the hidden layer and ways of reducing interference by introducing certain architectural constraints in the model will be the subject of the remainder of this section. We will first turn our attention to the analysis of another series of simulations performed by Ratcliff (1990) that may shed some light on this puzzling issue.

Network architecture and interference
The study by Ratcliff (1990) presents a series of simulations that investigate sequential interference in backpropagation networks. His main conclusion is that such networks cannot serve as good models for human memory. We fully agree with this conclusion. As we have argued in Parts I and II, modular networks (or similarly constrained architectures) may be better models of human

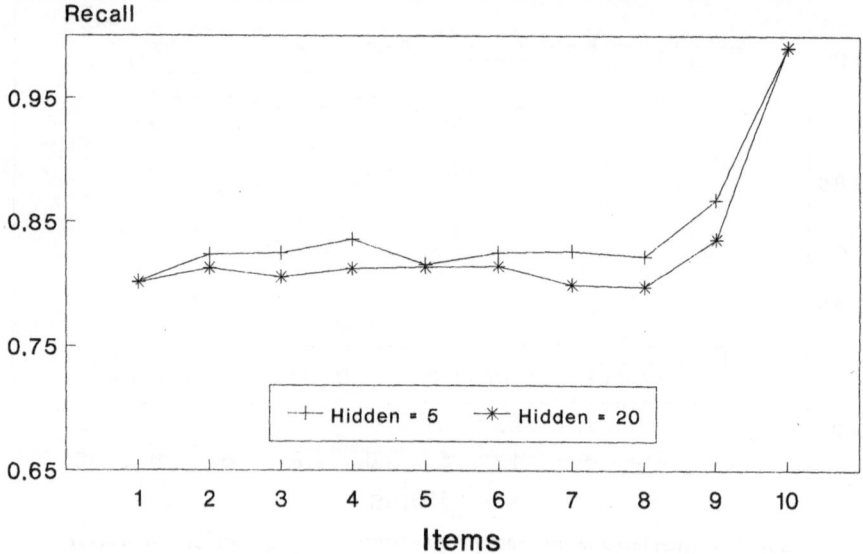

Figure 7.8 The negligible effect of increasing the size of the hidden layer on retroactive interference. See Figure 7.3 for an explanation. The results are averaged over 100 replications.

memory. Below we will analyse some variants of Ratcliff's studies that further reinforce this proposition.

In the first series of simulations Ratcliff (1990) trained a three-layer back-propagation network with four simple input patterns. The encoder scheme was used (e.g. auto-associative learning, see Rumelhart *et al.*, 1986). First, the network was trained to some criterion on patterns [1 0 0 0], [0 1 0 0] and [0 0 1 0]. Then the network was trained on the remaining pattern [0 0 0 1]. The criterion was that the difference between output and target should be below 0.06. After training, all four patterns were presented for recall (comparable with the studies above). Table 7.2 shows a replication of Ratcliff's simulations using three hidden nodes. Our replication fully confirmed his findings. It will be immediately clear from the table that sequential learning of the fourth item causes massive interference in the previous three. What is the reason for this interference?

Table 7.2 Replication of Ratcliff's (1990) results on interference with sequential learning using backpropagation. See text for an explanation.

| | Activation values | | | | | | |
| | Output layer response | | | | Hidden layer response | | |
	1	2	3	4	1	2	3
Training set:							
1 0 0 0	0.94	0.04	0.05	0.02	0.90	0.05	0.91
0 1 0 0	0.05	0.94	0.05	0.02	0.05	0.90	0.89
0 0 1 0	0.05	0.05	0.94	0.02	0.91	0.90	0.05
0 0 0 1	(additional training)						
Testing:							
1 0 0 0	0.73	0.00	0.00	0.92	0.90	0.05	0.91
0 1 0 0	0.00	0.67	0.00	0.93	0.05	0.90	0.89
0 0 1 0	0.00	0.00	0.71	0.93	0.91	0.90	0.05
0 0 0 1	0.04	0.04	0.04	0.94	0.65	0.68	0.67

| | Weight values | | | | | | |
| | Hidden → output | | | | Input → hidden | | |
	1	2	3	4	1	2	3
Before	1.66	−5.14	1.73	−2.09	2.18	−2.92	2.28
additional	−5.15	1.73	1.62	−2.09	−2.99	2.21	2.13
training	1.69	1.61	−5.14	−2.14	2.32	2.23	−2.87
					−0.22	0.18	0.05
After	0.74	−6.18	0.76	1.28	2.18	−2.92	2.28
additional	−6.13	0.60	0.58	1.50	−3.00	2.21	2.13
training	0.73	0.52	−6.16	1.36	2.32	2.23	−2.88
					0.60	0.78	0.70

Ratcliff argues that backpropagation primarily satisfies the constraints of the last items trained. If a model has first been trained on a block of items A, the solution found for A will form the starting point for the next block B. The algorithm will satisfy the constraints of the last block. Although one can hardly disagree with this argument, it does not satisfactorily explain the mechanism that underlies the massive interference. One might just as well suppose that the solution for B is found while the constraints for A are simultaneously being satisfied; the same would be found if A and B were trained in a single batch. We propose that this interference can be explained only with reference to the overlap of representations in the hidden layer of consecutive patterns. A similar explanation for sequential interference has recently been put forward by French (1991). Before discussing this proposition, let us first investigate more closely the replication results given in Table 7.2.

In general, the backpropagation algorithm has a tendency to find extreme values (i.e. near zero or one) for the hidden layer representation (Rumelhart et al., 1986, p. 337). In Table 7.2, we see that indeed after training on the first three patterns the activation values are either around 0.05 or around 0.90. Before any training had taken place, these values were centred around 0.50. This is due to the initialization with random weight values in the range $[-0.3, 0.3]$ (i.e. with an average value of zero, giving an average activation value of 0.5). Before learning, the fourth pattern's 'spontaneous representation' also in the hidden layer consisted entirely of values around 0.50. Training on the fourth pattern has slightly increased these values to around 0.67. The corresponding weights have increased, as can be seen in Table 7.2 (input → hidden). Note that all other weights from the input to the hidden layer have remained unchanged. This is due to the learning rule in backpropagation that says that no weight change is to take place if the pre-synaptic activation equals zero (cf. the zeros in the patterns). In the table, it can also be seen that after training on the fourth pattern, the values of the weights from the hidden layer to output node 4 have changed from -2.09, -2.09 and -2.14 to 1.28, 1.50 and 1.36, respectively. All other weight values from the hidden layer to the output layer have been decreased in value.

In short, training on the fourth pattern has resulted in a set of weights that is optimal for this pattern, but to obtain this set of optimal values the weight values from the hidden layer to the output layer especially had to be strongly modified. And because (i) the hidden layer representations of the first three patterns have been left untouched, and (ii) these representations show considerable overlap with the fourth pattern's hidden layer representation, the first three patterns tend to reproduce the last pattern. The algorithm seems to work by recoding the input patterns to 'random bit vectors' in the hidden layer. But it only discriminates well between patterns *within* a single block. 'Random bit coding' of two *consecutive* blocks appears to take place entirely independently (i.e. on this particular task).

Consecutively learned blocks of patterns will interfere to the extent that their between-block hidden layer representations overlap. This appears to be a chance process. If, for example, we approximate this process by simplifying the hidden node values to binary activations, then it will be immediately clear that the expected overlap between the fourth pattern and any of the earlier patterns is 1.5 (i.e. half of the hidden nodes will overlap). The chance that a certain node has, say, the value 1 for both the first pattern and any given earlier pattern is, of course, 0.5. The overlap, thus measured, follows a binomial distribution, with a mean equal to half the size of the hidden layer.

We can now easily understand why increasing the hidden layer will not decrease interference. The reason for this is that this operation does not reduce the overlap. On the contrary, while it does not affect relative overlap, it may increase absolute overlap (measured as the number of coinciding hidden node values between two patterns). This may also explain the slight decrease in performance of the fourfold increased hidden layer in Figure 7.8. For the case of Ratcliff's simulations, we have found that an increase in the number of hidden units indeed appears to have a negligible effect on the interference. Ratcliff (1990) reports a variant of this, where he added an extra hidden unit for the fourth pattern to be learned. With training on the fourth pattern he allowed the algorithm to modify all the weights. This still caused considerable interference (see his Table 3). We may explain this result by arguing that the overlap in the hidden layer will be reduced, but not eliminated. Even when we would only allow weights from the newly added hidden node to be changed, we would still expect interference, as has indeed been reported by Ratcliff (1990, Table 4).

In conclusion, the results of the first series of simulations in Ratcliff (1990) can be explained in terms of overlap in the hidden layer. If we want to reduce interference, we must therefore direct our attention to a reduction of hidden layer overlap.

A similar conclusion has recently been drawn by French (1991). He describes an extension, or rather a constrained variant, of backpropagation that is able to reduce the overlap by simply allowing only k nodes in the hidden layer to be active. He calls this a semi-distributed representation, thereby referring to other authors that have described models employing such representations (Kanerva, 1988; Kruschke, 1990). We would add to this that all modular neural networks use semi-distributed representations. Distribution is optimized if a competitive wiring scheme is used as in the CALM networks described in this work. Ratcliff's simulations could trivially be implemented by a CALM network with the architecture [Input module] → [Hidden CALM module] → [Output CALM module]. All of the modules would have to be of size 4 (see also Figure 3.7 for a similar architecture and training method). As was shown in Part I, CALM has no difficulty in discriminating the patterns used in Ratcliff's simulations. In the 'hidden CALM module', owing to the learning rule used, an orthogonal representation will emerge. The model will thus perform perfectly, without any interference, no matter what sequence of pattern presentation is used. Thus, modular neural networks do not suffer from catastrophic interference. This holds true as long as none of the module's capacity is exceeded. In that case, some representations will, of course, coincide, for example when a number of hidden CALM modules with sizes of 2 and 3 are used.

French (1991) calls his method 'activation sharpening'. It allows a back-propagation network to develop semi-distributed representations in the hidden layer. He argues that, 'if the system could evolve representations with a few highly activated nodes, rather than many nodes with average activation levels, this would reduce the amount of activation overlap among representations' (French, 1991, p. 4). Nodes are sharpened by increasing the activation values with $\alpha(1 - A_{old})$, the other nodes are simply decreased by a value αA_{old}, where α, the sharpening factor, is a small constant value, and A_{old} is the activation value before sharpening. The method of k-node sharpening works by (i) finding the k most activated nodes in the hidden layer (first perform a forward pass), (ii) sharpening these nodes, (iii) using the difference between the old value and the sharpened value as error, (iv) adjusting the weights to the hidden layer according to these errors, and (v) executing the backpropagation algorithm as usual for a single pattern iteration, at each pattern presentation (i.e. iteration).

Simulations by French (1991) indicate that sharpening does reduce sequential interference, and that this effect can be related to a reduced amount of overlap in the hidden layer. Neither the simulations, nor the patterns are described in any detail, but he appears to measure recall in terms of 'memory refresh time' (cf. the discussion of savings above). A block of 11 items is trained using an 8–8–8 backpropagation network, probably on an auto-associative task. The learning rate reported is 0.2, the momentum is 0.9, and the sharpening factor is 0.2. After training, a single new item is presented for learning. The network is then tested for savings. Memory refresh time is minimal for the case of two-node sharpening. For k-node sharpening with k greater than four, the variant algorithm performs no better than backpropagation.

We have repeated the simulation of Table 7.2 using k-sharpening. The results are given in Table 7.3 for $k = 1$ and a hidden layer of four nodes. Interference is strongly reduced, although it is not entirely absent. This is probably due to some remaining overlap of the hidden layer representations. The fact that this particular simulation is successful does not imply that the k-sharpening always works. Two situations may occur in which the method fails. Firstly, the hidden layer representations in the first block of patterns may by chance coincide. In this case, the algorithm is usually not able to develop discriminating hidden layer representations and the simulation fails to reach the convergence criterion. In practice, this is especially likely to happen when k is greater than 1. For $k = 1$, the algorithm seems to be able to find orthogonal representations most of the time. But for higher k, the process appears to be entirely stochastic, and the chances of no convergence – which are surprisingly high – can be predicted in advance (see below). Secondly, the hidden layer representation of the fourth pattern may overlap with any of the patterns

Table 7.3 Replication of Ratcliff's (1990) results on interference with sequential learning, using k-sharpening with $k = 1$ (French, 1991). See text for an explanation. The learning rate was 0.2, the momentum was 0.9, and the sharpening factor was 0.8. Note that the hidden layer response of the first three patterns is not affected by training the additional pattern. Therefore, only final representations are shown.

				Activation values			
				Output layer response			
				1	2	3	4
Training:							
1	0	0	0	0.94	0.06	0.06	0.06
0	1	0	0	0.06	0.94	0.06	0.05
0	0	1	0	0.06	0.06	0.94	0.06
0	0	0	1	(additional training)			
Testing:							
1	0	0	0	0.90	0.02	0.02	0.23
0	1	0	0	0.03	0.87	0.02	0.22
0	0	1	0	0.03	0.02	0.87	0.22
0	0	0	1	0.05	0.06	0.06	0.97
				Hidden layer response			
				1	2	3	4
Testing:							
1	0	0	0	0.05	0.05	0.06	0.97
0	1	0	0	0.97	0.05	0.07	0.05
0	0	1	0	0.05	0.97	0.07	0.05
0	0	0	1	0.05	0.04	0.96	0.05

learned earlier. This situation is shown in Table 7.4, where the simulation of Table 7.3 is repeated with a network of eight hidden nodes. As can be seen, for the first and third pattern interference is strongly reduced, but the second pattern is completely overwritten by the fourth pattern. This failure of recall can be attributed to coinciding hidden layer representations (printed in bold type).

In Table 7.5, results for the same network are shown, but with $k = 2$. The first pattern is partially overwritten by the fourth. This can also be explained by reference to the hidden layer, where one of the positions of the two active nodes coincides. Patterns 2 and 3 are reproduced much better, but they show more interference than the patterns of Table 7.4. This is probably due to crosstalk from the six inactive nodes in the hidden layer that still have

Table 7.4 Replication of Ratcliff's (1990) results on interference with sequential learning, using k-sharpening (French, 1991). See text and Table 7.3 for an explanation. The difference with Table 7.3 is that a larger hidden layer has been used. Overlapping representations have been printed in bold type.

				Activation values			
				Output layer response			
				1	2	3	4
Training:							
1	0	0	0	0.95	0.06	0.05	0.06
0	1	0	0	0.05	0.96	0.05	0.05
0	0	1	0	0.05	0.05	0.95	0.06
0	0	0	1	(additional training)			
Testing:							
1	0	0	0	0.94	0.02	0.02	0.28
0	1	0	0	0.02	0.05	0.02	0.98
0	0	1	0	0.04	0.00	0.94	0.28
0	0	0	1	0.02	0.06	0.03	0.98

				Hidden layer response							
				1	2	3	4	5	6	7	8
Testing:											
1	0	0	0	0.05	0.06	0.05	0.04	0.97	0.07	0.05	0.04
0	1	0	0	0.05	0.04	0.06	**0.98**	0.04	0.04	0.05	0.04
0	0	1	0	0.06	0.05	0.05	0.04	0.03	0.05	0.06	0.97
0	0	0	1	0.03	0.04	0.03	**0.97**	0.04	0.04	0.03	0.04

activation values of around 0.08. A conclusion that can be drawn from this simulation is that even overlap of a single node may cause considerable interference.

From the above simulations we may draw the general conclusion that overlap of hidden layer representations is a major factor in retroactive interference. Interference can be reduced by reducing representation overlap between patterns that are learned consecutively. We will conclude Section 7.4 with an analysis of the expected representation overlap and its consequences.

Analysis of hidden layer overlap

For the case of k-node sharpening, we can easily derive the probability density function of the overlap. Suppose we have a hidden layer of size n in which exactly k nodes are active (i.e. k-node sharpening). If for two patterns at exactly c positions both patterns have active nodes, we say that the extent of the overlap is equal to c. The probability density function for c is given by

Table 7.5 Replication of Ratcliff's (1990) results on interference with sequential learning, using k-sharpening (French, 1991). See text and Table 7.3 for an explanation. The difference with Table 7.3 is that a larger hidden layer has been used, and $k = 2$. Overlapping representations have been printed in bold type.

					Activation values					
					Output layer response					
					1	2	3	4		
Training:										
1	0	0	0		0.95	0.05	0.06	0.05		
0	1	0	0		0.06	0.95	0.05	0.05		
0	0	1	0		0.05	0.06	0.95	0.04		
0	0	0	1		(additional training)					
Testing:										
1	0	0	0		0.51	0.01	0.02	0.86		
0	1	0	0		0.00	0.88	0.02	0.49		
0	0	1	0		0.00	0.02	0.89	0.47		
0	0	0	1		0.06	0.04	0.04	0.99		

				Hidden layer response							
				1	2	3	4	5	6	7	8
Testing:											
1	0	0	0	0.09	0.11	**0.93**	0.09	0.09	0.11	0.09	0.94
0	1	0	0	0.11	0.11	0.09	0.93	0.10	0.93	0.10	0.07
0	0	1	0	0.14	0.11	0.08	0.08	0.93	0.08	0.93	0.10
0	0	0	1	0.07	0.94	**0.92**	0.08	0.08	0.07	0.07	0.05

the function $Q(c)$:

$$Q(c) = \binom{k}{c}\binom{n-k}{k-c} \Big/ \binom{n}{k} \qquad (7.14)$$

The derivation is as follows. Consider two patterns A and B, each with k active nodes and an overlap of c. This means that at c positions of the k active nodes in A, pattern B also must have active nodes (and inactive nodes at the remaining positions). There are $\binom{k}{c}$ ways in which these c nodes in B can be permuted. For the remaining $n - k$ positions in A (with inactive nodes) there remain exactly $k - c$ active nodes in B, which gives $\binom{n-k}{k-c}$ permutations. Because both permutations are independent, they may be multiplied to give the total number of permutations with overlap c. To arrive at the probability density, this number must be divided by the total number of all permutations possible with k active nodes in a layer of size n. This results in equation (7.14).

If the overlap of two patterns is greater than one, the first pattern is almost completely overwritten by the second. We are, therefore, most interested in the chances of an overlap of zero or one. In Figure 7.9, these probabilities are

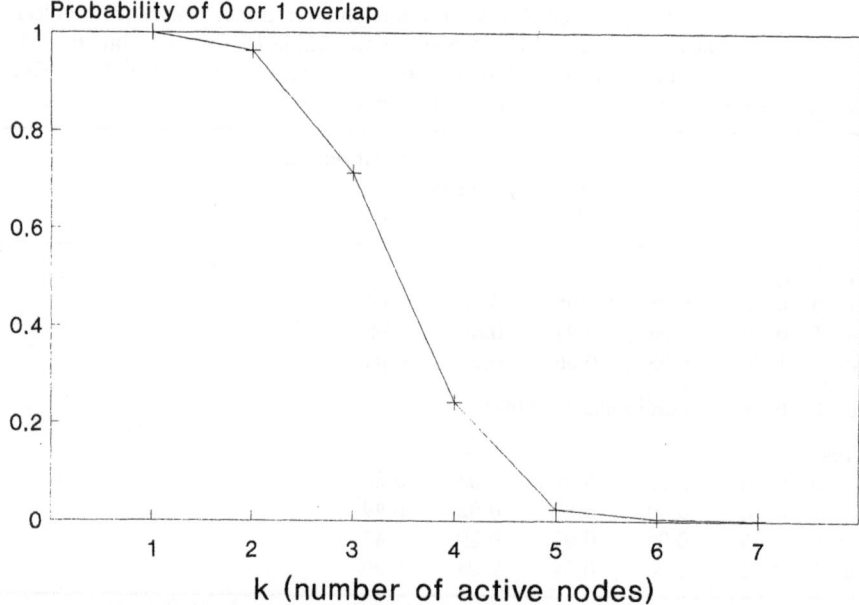

Figure 7.9 Probability of an overlap of either zero or one in a hidden layer of eight nodes. Values for increasing numbers of active nodes, k, have been plotted.

plotted for increasing values of k in a hidden layer of size 8. From the figure it will be clear that a value of $k = 1$ gives the least overlap. Also from the figure it can be seen that for $k > 4$, the probability of overlap of at least one node becomes very high, which may explain the results by French (1991, discussed above), who found that for $k > 4$ the model did not do much better than normal backpropagation.

With equation (7.14) we may calculate the probability that a new pattern to be learned has an overlap of c with any given pattern learned earlier. In a similar fashion we may derive the probability that any two (or more) P patterns have exactly the same representation (i.e. maximal overlap). In such a case, a block of these patterns will fail to converge, as was noted above. This problem is similar to the well-known 'birthday problem' that deals with the question: 'How many people in the room have the same birthday?' The chances of this happening are deceptively high. The 'number of people' in our case is P, and the 'number of birthdays' (i.e. 365 in normal years) corresponds to the 'number of ways two patterns can form maximal overlap'. The latter number is equal to the number of representations, which is simply equal to the number of permutations of k in n (see above).

The derivation of the solution for the 'birthday problem' is straightforward. It is most easily approached by first calculating the probability that everybody has their birthday at a different date. The first person can freely 'choose' his or her birthday. The second person has one less day to choose from, the third person two less days, etc. All of these chances must be multiplied, which gives

$$1 \cdot \left(\frac{365 - 1}{365}\right) \cdot \left(\frac{365 - 2}{365}\right) \cdots \left(\frac{365 - (P-1)}{365}\right) \qquad (7.15)$$

where P is the number of persons, or patterns, as the case may be. In general, the chance of no pattern overlap is:

$$\prod_{i=0}^{P-1} \binom{M-i}{M} = \frac{M!}{(M-P)! M^P} \quad \text{with} \quad M = \binom{n}{k} \qquad (7.16)$$

For example, with a hidden layer of size 8 ($n = 8$), $k = 2$ and four patterns ($P = 4$), we find that M, the number of permutations, is 28. Filling in these figures in equation (7.16), we find that the probability that no patterns have maximal overlap is 0.80. This means that approximately one out of five simulations with these parameters will not converge. For 10 patterns, equation (7.16) gives 0.16. This implies that about seven out of eight simulations will not converge. It can be shown that for any value of k, including more than three patterns in the block will increase the chances of no convergence above 0.10. Including more than 10 patterns will increase this probability above 0.50. Thus, it appears that even for modest values convergence problems may be expected. (This makes the birthday problem interesting; with 35 persons already present in the room, the chances of two or more persons having the same birthday are high.)

7.5 Conclusions

This work advocates a modular approach to neural networks. In Part I, the CALM model was introduced to function as a basic building block for modular networks. In Part II, it was shown how neural networks assembled from CALM modules can be applied to psychological modelling and to practical applications such as pattern recognition. It was also shown how genetic algorithms can be used automatically to design the architecture of a modular neural network. The overall conclusion of Appendices B1–B5 will be that modularity is especially important for the feasible implementations of neural networks in parallel hardware. For virtual implementations modular neural networks have better speed-up and efficiency characteristics. Modularity also offers an important solution to the problem of scalable designs for dedicated neural network hardware.

In this part of the work, several of the issues introduced in Part I were revisited. It was shown that the strict pairing of R-nodes and V-nodes, which is not biologically plausible, is not a requirement for CALM. On the contrary, it was concluded that loosening this restriction may offer an alternative explanation for the emergence of topological organization. It was, furthermore, argued that the categorization and learning mechanism in CALM causes input patterns to become decorrelated in a 'negative filter' that is maintained through novelty-dependent learning. It was also concluded that from both a biological and a computational point of view the maximum storage counter-argument against modularity cannot be defended.

From a review of several studies investigating the effects of constrained architectures, such as modular networks, it was concluded that limiting the structure of neural networks may improve several aspects of their behaviour, such as quality of solution and learning–convergence time. In particular, based on the results from formal learning theory, the significant conclusion was drawn that constrained architectures will have better generalization behaviour.

A detailed analysis of replications of simulations by Ratcliff (1990) led to the conclusion that retroactive interference in backpropagation networks can be explained entirely in terms of overlap of representations in the hidden layer. Interference can be reduced by reducing representation overlap between patterns that are learned consecutively. Several methods to attain this were tested, all based on 'sharpening' the hidden layer representations.

From the simulations and from the analysis, we may conclude that the method of sharpening does reduce overlap, and hence may reduce inter-ference, but that it has two serious drawbacks. Firstly, it may fail to converge, even for small problems. Secondly, chances of overlap between a new pattern and any given old pattern remain fairly large. Although the introduction of semi-distributed representations in backpropagation may lead to behaviour that is psychologically more plausible in certain respects, it is still not suited as a model for human memory.

As remarked above, CALM would reduce interference to zero. The reason for this is that a CALM module is able to discriminate between subsequent items, as long as they are sufficiently different. The CALM module categorizes these items on different nodes. Another way of looking at this is by saying that a CALM module is able to develop *orthogonal* internal representations, even with patterns learned consecutively. Backpropagation fully lacks this ability, except when items are learned in a single block. This is the main reason why even a substantial increase in the size of the hidden layer has a negligible effect on catastrophic interference. We conclude that one of the important aspects of unsupervised learning lies in the ability to develop distinct internal represen-tations independently of the manner of presentation.

Appendix A

The parameter set shown in Table A.1 has been used throughout all simulations. The largest part of the parameters are fixed-weight values used mainly for the intramodular connections. The remaining parameters refer to the learning and activation rule, and to the convergence criteria.

Many of the parameters can be derived from qualitative considerations (e.g. the absolute value of the 'high' weight should be larger than the 'low' weight to have no A-node activation in the absence of competition). In fact, global range values can be deduced for many parameters from these considerations. Although the parameters may be optimized for certain specific practical tasks (see Chapter 6), in none of the simulations of the psychological experiments reported in this book did we strive for a precise quantitative fit. The exact value of the parameters does not seem to be very critical and indeed very large parameter variations have been shown to produce similar qualitative behaviour of a CALM module. The present parameter set has been selected by computer simulations to produce the quickest convergence in a single CALM module. The number of parameters appears to be higher than the number of degrees of freedom in the model. Shifts in one parameter can mostly be compensated by shifts in other parameters (e.g. a change in AE weight can be counteracted by a change in strange weights (or ER weights) and $W_{\mu E}$ weights). Adherence to the architectural scheme, which was proposed for other reasons, however, forces us to use this large number of parameters.

To count convergences, threshold values for winning and losing nodes have to be specified. In all simulations reported a convergence took place when the winning node had an activation value above the high threshold c_h and all other activations were below the low threshold c_l. These values have no influence on the actual functioning of the networks; they just serve as a criterion for counting output productions.

Table A.1 Weight and parameter values used throughout this book. The ELAN weights are only used in the ELAN models of Chapter 4. For certain weights and parameters, ELAN models have slightly different values: down weight, -1.0; high weight, -2.0; interweight, 0.5; k, 0.25; L, 2.0; c_h, 0.1; c_l 0.05.

	Description	Value
Weights:		
Up	Connects R-node to its matched V-node	0.5
Down	Connects V-node to its matched R-node	-1.2
Cross	Connects V-node to all non-matched R-nodes	-10.0
Flat	Interconnects V-nodes	-1.0
High	Connects V-nodes to A-node	-0.6
Low	Connects R-nodes to A-node	0.4
AE	Connects A-node to E-node	1.0
Strange (or ER)	Connects E-node to R-nodes	0.5
Inter	Initial value of learning weights	0.6
ELAN weights:		
AV	Connects A-node to V-node in the output module	1.0
VA	Connects additional V-node in CALM to A-node	3.0
Reset	Connects output nodes to the additional V-node	2.0
VV	Connects additional V-nodes of two modules	0.4
Output	Connects R-nodes to output nodes	2.0
Parameters:		
k	Decay of activation in an iteration	0.05
K	Maximum value of learning weights	1.0
L	Learning competition	1.0
d	Base rate of learning	0.005
$W_{\mu E}$	Virtual weight between E-node and learning rate	0.05
c_h	High convergence threshold	0.001
c_l	Low convergence threshold	0.0001

Appendix B1

Hardware and Software for Neural Networks

B1.1 The importance of implementation

Efficient implementation of a model can be of crucial importance for both modelling studies and practical applications. In particular, neural network models require extensive computational facilities. So far, most simulation studies have been carried out on conventional computers, and a number of commercial applications have been developed for serial machines enhanced with special coprocessor boards. Operational applications that run on parallel neurocomputing hardware are still rare, although a few commercial systems have recently been developed (e.g. LeCun *et al.*, 1990b). Even with small learning networks, training – and to a lesser extent executing – networks on a conventional machine can be extremely time consuming (days or weeks of CPU time). And as models are growing more complex, and applications become more sophisticated, fast hardware implementations will be a necessity rather than a luxury. One of the major conclusions of the DARPA (1988, p. 52) report stated this as follows:

> Hardware capabilities are limiting the development of important neural network applications. It is clear that if researchers are not provided with improved simulation and implementation capabilities, the field of neural networks will once again drift off into the wilderness.

The greatest limiting factor of conventional hardware is *speed*, and much of Appendices B1–B5 will concentrate on ways to improve the speed of neural network implementations.

Even if a connectionist model is suited for implementation in parallel hardware and even if such hardware is available, the interaction with the model and the integration of the computational resources can only be accomplished if supporting software is available. The right neural network simulation environment is as important as a fast machine. Anderson and Rosenfeld (1988, p. 714) remark in their assessment of the field of connectionism:

> The software and applications will be by far the most difficult, important and painful aspect of practical applications ... software and applications will be many times more difficult than hardware.

The major cause for this difficulty is that parallel software lags far behind the hardware. Most software, such as compilers, only supports a straightforward mapping of

155

instructions and data onto the parallel architecture of a specific machine. Sometimes, one extra layer of abstraction is added so that certain scheduling and synchronization instructions are handled automatically. In many cases, even such an extra layer is lacking and the programmer is left to develop his or her own communication and routing functions. Different ways to alleviate this problem are being explored.

The ideal situation seems to be the automatic conversion of existing programs, say in C or Pascal, into parallel versions. For a limited number of applications this is indeed possible, and a great deal of work is being done on so-called parallelizing compilers. Methods of automatically extracting the implicit parallelism in serial programs, however, may have serious limitations. A more important drawback is that automatic conversion systems cannot invent new parallel algorithms (Gelernter, 1987).

The only fundamental solution seems to be luring programmers into 'thinking parallel' by offering programming languages or environments that invite such solutions. This must not be confused with offering support at the level of the operating system, but rather in the language itself (Gelernter, 1988). In particular, the language must allow for a specification of the *logically* concurrent processes. Such a specification is inherently hardware independent. *Physical* concurrency is achieved by converting the program into a network of instructions in accordance with the available hardware. The conversion may be optimized by using load-balancing algorithms. As we will argue in Appendix B5, a unified approach to modelling, user interaction, and system integration of computational resources may be the best way to proceed with this difficult problem (see also Hudson *et al.*, 1990; Hudson and Murre, 1991; Murre and Kleynenberg, 1991).

Because methods are being discussed here for implementing neural networks in different types of parallel hardware, as a short introduction the next section will address several approaches to parallelizing neural networks. It will be followed by a section briefly reviewing hardware for neural networks, and stressing its particular importance for the field of connectionism. In a final section, a broad classification of neural network implementations will be presented that serves as an organizing framework throughout these appendices.

B1.2 Parallelizing neural networks

The degree of parallelization that can be obtained with a neural network implementation is limited only by the inherent parallelism of the model and the nature of the available hardware. If the hardware fits exactly the needs of a connectionist model a *physical implementation* is possible: every node is implemented by a processor, and all model (node-to-node) connections can be mapped onto hardwired (processor-to-processor) connections. This situation will most certainly occur only with dedicated neurocomputers, in particular if they are of the analog type (see Section B1.4, and Appendix B4). More often it will be possible to implement each node by one processor, but impossible to map all of the model connections onto the hardwired connections. This implies that several model connections must be mapped onto one hardwired connection; a single wire must transmit the signals of the connected nodes consecutively. This technique is called *multiplexing*. An implementation where each node is implemented by a processor, but connections are multiplexed, will be called *hybrid*. If the nodes are multiplexed as well, we will speak of a virtual implementation. Appendices B1–B5 are organized around this principal tripartition. In Appendices B2–B4, examples will be given of virtual, hybrid and physical implementations of CALM, respectively.

Notwithstanding the greater potential of hybrid and analog neurocomputers, the current world speed record is still in the hands of a 'conventional' virtual neural network implementation (Singer, 1990). In the past 5 years, different types of parallel computers have become commercially available. Many institutes have access to transputer networks (see Appendix B2), N-cubes, or some type of vector processor. A few fortunate organizations even possess large parallel computers such as the Connection Machine (Hillis, 1985, 1987). Much effort is being spent on the development of even larger and faster systems. Some of these are intended for special purposes, while others must serve as general purpose multiprocessors. The term multiprocessor as used here may be criticized because in the following appendices it will usually refer to systems consisting of several independent processors, where each of these processors has its own memory (e.g. a transputer). Some authors (e.g. Stein, 1988) propose to call such systems multi*computers* to distinguish them from shared-memory multi*processors*. These latter systems typically suffer from the 'von Neumann bottleneck', because if a processor wants access to the shared memory it must wait for permission from all the others. Similarly, all other processors must wait until the memory transfer has ended. Adding more and more processors will eventually lead to unacceptably long waiting times. Multicomputers on the other hand communicate through direct or indirect links. In these systems the 'von Neumann bottleneck' is generally absent, but other problems arise as we will see below.

The most straightforward way to parallelize a neural network on a multiprocessor is by multiplexing nodes over the available processors. Assuming that we have a neural network with m nodes and a transputer network (or N-cube or similar machine) with n processors, where n is smaller that m, each processor must implement several (e.g. m/n) nodes. For such an implementation to be efficient, interprocessor communication should not take too much time. Furthermore, the neural network model must have synchronous temporal evolution (see Section B1.4). If the model uses an asynchronous activation rule (e.g. a Hopfield (1982) network), parallelization is impossible, because by definition only one node can be processed at a time, which implies sequential processing of the nodes. Models with irregular temporal evolution (see Section B1.4) impose a specific order in which certain network objects should be processed. In backpropagation, for instance, in the up-sweep activations are updated sequentially over the layers: first the input layer, then hidden layer 1, hidden layer 2, ..., and finally the output layer. Then, in the down-sweep update processes are executed in reverse. In these cases, multiplexing can only be successfully achieved within synchronously updated collections of nodes (i.e. within layers or modules). Appendix B2 presents an analysis of virtual implementations with synchronous temporal evolution. In Appendix B5, a general framework will be outlined for dealing with networks with irregular temporal evolution, which is based on formal theories of concurrent processes.

When the number of processors n is (much) larger than the number of nodes m, each *node* may be assigned to a *subset S* of the available processors. This situation is not likely with (expensive) transputer networks, but it may occur in systems based on dedicated neurochips with large numbers of atomic processors tailored to the task of processing neural networks. In such systems, the connections to a certain node may be multiplexed over the processors in the assigned subset S. For instance, in most neural networks for the execution of the activation rule it is necessary to calculate the inner product between the incoming activations and the corresponding weights. If a tree-like processor topology were available with $2z - 1$ atomic processors for the weights to a single node, then at each of the z leaves a number of activations could be multiplied with the weights. By adding the values in the leaves in a tree-like manner the inner product would appear in the root after log(z) of such intermediate addition steps (see Figure B1.1). At the root a sigmoid function (or other activation rule) could be applied to the total input activation and the resulting new activation value could be transmitted back in log(z) steps to the leaves. There, the learning rule could be applied to each of the weights implemented at that leaf. Suppose the number of incoming weights to a node is 1024 and 2047 atomic processors are available to implement these. Then, if interprocessor data-transfer time were negligible, the time to add up the weighted activations would be log(1024), or 10 time units, instead of 1024, which is a speed-up of more than a factor of 100. In addition to this, the multiplication of the activations and weights as well as the application of a learning rule to each weight would take only 1/1024 of the time compared with serial execution. The latter factor will usually contribute significantly to the overall reduction of processing time per cycle. For instance, in the BSP400 (see Appendix B4) it was found that calculating the inner product of weights and activations and then application of the activation and learning rules took by far the most time (more than 90%). In physical implementations the above sketched method may be hardwired. This would be possible, because tree topologies have a limited degree of connectivity.

As remarked above, neural network models with an irregular temporal evolution, such as backpropagation, may pose problems if parallelization is attempted by multiplexing over the activations. With backpropagation, however, one can have recourse to the possibility of *batch learning*. With batch learning the weights are updated only after presentation of an entire batch of patterns, rather than after each individual pattern presentation (the latter method is sometimes called *stimulus learning*). Indeed, in the formulation of Rumelhart *et al.* (1986), backpropagation is derived as a method of gradient descent in error space, where the total error is taken *over the patterns*. Thus, batch learning is the 'correct' method. If this method is used, the implementation can be parallelized over all patterns. Each processor stores a copy of the entire network, and each of the processors is assigned an equal number of training patterns from the batch. In other words, the batch of patterns is divided over the processors. After all patterns have been processed, the updated weight changes are exchanged among the processors and added, resulting for each weight in a weight change calculated over all

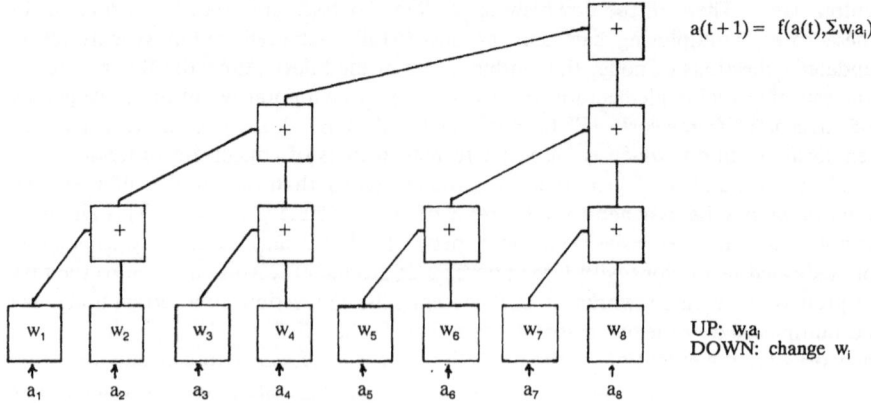

Figure B1.1 Dividing incoming weights to a node over a number of processors in a tree topology.

the patterns. Finally, each processor updates the weights in its copy of the network, so that overall integrity is preserved. This process is repeated at each cycle. Singer (1990) uses this technique to achieve the extremely high speed of 1.3 giga connections per second (GCPS). He uses a Connection Machine with 65 536 processors to achieve this. Witbrock and Zagha (1989) reach 0.9 GCPS using a similar method on an experimental machine called the GF11 with 356 processors. They, furthermore, use a tree-like processor interconnection topology in the GF11 to add the accumulated weight changes, as well as a combination of table lookup and linear interpolation to calculate the sigmoid activation function. The tree-like interconnection architecture somewhat resembles the method of parallelizing over weights outlined above. Singer (1990) does not report many details, but he does remark that several low-level improvements are still possible in his implementation that may lead to a further speed increase of a factor of five to 10. This would bring implementations well into the GCPS range.

Neural networks with irregular time evolution often allow for specific types of parallelization. For backpropagation, Singer (1990) remarks that the up- and down-sweeps may be pipelined over subsequent patterns. While the errors (deltas) of a pattern A are being backpropagated from the output to the hidden layer(s) (down-sweep) the activations for the next pattern B may already be passed through the weights from the input layer to the hidden layer (first hidden layer in the up-sweep). The sigmoid function can only be applied to the inputs after the weight changes due to pattern A have been executed. Some dedicated neurocomputers have been developed with the explicit aim of implementing neural networks with irregular temporal evolution, for example to implement ART 1 (Grossberg, 1976; Carpenter and Grossberg, 1987, 1988). This model uses a non-local activation rule to establish which is the winning (and therefore to be activated) node in the upper, F2 layer. This winner-take-all mechanism seems to be better suited to a (deep) pipelined architecture than to a fully parallel multiprocessor architecture. In a dedicated, pipelined implementation of ART 1 by Rao et al. (1990), all incoming patterns are subsequently compared with all nodes. Each pattern 'remembers' which is the best fitting node so far. As soon as it reaches the last node, the pattern may conclude which node it will be assigned to. A fully parallel, physical implementation of this mechanism will run into problems, because any given node cannot possibly 'know' whether any other node receives more total input activation.

An interesting approach to parallelization is described by Fukuda et al. (1990). Both connections and nodes are mapped onto a torus of virtual processors. Rows and columns in the matrix describing the virtual processor structure on the torus are permuted to balance the load within each region of the matrix. The matrix is then partitioned into n submatrices, each of which is assigned to one of the n processors available (i.e. 16 transputers). Although this approach does indeed effectively balance processing load it does not address the question of interprocess communication. The authors are not very clear on this point, and only remark that overhead time for communication can be neglected. This does not agree with the findings described in Appendix B2.

One final approach to parallelization may be mentioned here, although it hardly deserves this name: running m independent simulations simultaneously on m processors. This technique is often used for the implementation of genetic algorithms (see Section 6.5.3). It can easily be applied and it can be highly effective.

Summarizing the above, parallelization of neural networks can be achieved by parallelizing over nodes, connections, patterns or simulations, depending on the type of model and available hardware. For some modules, full parallelization is not possible and pipelining must be used. In the next section, some remarks will be made on the motivation for the hardware implementations of neural networks, followed by a broad classification that serves as a general framework for the implementations discussed here.

B1.3 Motivation for hardware implementations

Probably the oldest recorded attempt to develop a hardware implementation of a neural network model was by Marvin Minsky and Dean Edmonds in 1951 (described in Rumelhart and Zipser, 1985). In the spirit of Hebb's (1949) ideas they built a learning machine that consisted of 300 tubes and many motors. They seem to have been motivated mainly by sheer fascination with learning machines and even today this may well be one of the hidden forces behind much of neural network research. Two other pioneering projects with learning machines (also neural network implementations) are mentioned in Swaine (1989): MINOS I and MINOS II. These were projects contracted by the US Army. The machines were built from 1958 to 1967. It is interesting that, in 1960, a routine test on the MINOS I, a three-layer machine, was to learn the EXOR. The army contract may have been one of the reasons why relatively little has been published on the precise nature of the hardware developed and the motivation for undertaking these projects. We may assume that the army saw much potential in a very fast machine that 'can take a few bullets and still run'. A neurocomputer is extremely fast on certain tasks and, if well designed, should be able to exhibit that general characteristic of neural networks called 'graceful degradation'. Apart from robustness, speed and sheer fascination with learning machines, there exist a number of other reasons why connectionist researchers have been involved with hardware implementations.

An important reason for some of the attempted hardware realizations of connectionist models is that problems with a model may emerge only during an actual implementation process. Problems may be due solely to the nature of the hardware, for instance faulty connections, unreliable components, overproduction of heat, etc. But most often an implementation clearly exposes the weaknesses and strengths of a model. For instance, only when it is attempted to implement a fully connected network model (e.g. a Hopfield (1982) model) in electronic hardware does the difficulty of interconnecting all nodes become fully appreciable. Recognizing this constraint, many research groups have opted for the implementation of networks with limited fan-in to the nodes. Implementation problems, in turn, may influence the community of connectionist researchers to pay more attention to certain aspects of neural network models. Limited fan-in of connections, for instance, invites the study of modular neural networks and in general stimulates the study of the behaviour of constrained network topologies. In Appendix B2, for example, it will be shown that, from an implementation point of view, it is extremely unlikely that a significant portion of the neurons in the brain are randomly interconnected.

Many researchers point out the great similarity between biological mechanisms and the characteristics of basic electronics components. The variety of non-linearities found with these components has a counterpart at the level of the neuron. As will be described in Appendix B4, such similarities have motivated some researchers to implement biological neural networks directly in analog hardware.

A general interest in parallel computers also contributes to the interest of the hardware implementations of neural networks. Conventional hardware is slowly approaching its theoretical limits. The application of novel materials and techniques, such as gallium arsenide (Brodsky, 1990) and X-ray lithography, may take us yet another step up the ladder of speed and memory capacity, but we are already very close to the boundaries. This insight has prompted a search for fundamentally different approaches to computer architectures, different, that is, from the descendants of the traditional von Neumann design. Among those approaches to computer design we find optical and electro-optical computers (Abu-Mostafa and Psaltis, 1987) and more 'exotic' forms such as molecular computers (Conrad, 1985). The inherent parallelism of neural networks makes them exceptionally well suited for implementation on these novel types of computers. Many experimental optical computers have been designed to implement neural networks (e.g. Treleaven, 1989).

Finally, the upsurge in parallel computer architectures has forced programmers to find ways of exploiting the possibilities offered by specific types of hardware. For many problems this is not an easy task. Usually, the programmer has to tailor critical parts of the program to the demands of the architecture. Sometimes, support is available in the form of so-called parallelizing compilers, but research in this area has not led to many generally applicable solutions. In some cases, the compiler paralyses, rather than parallelizes, the program (see Appendix B2) and interprocess communication must still be mapped out in full by the programmer. Furthermore, debugging such parallel programs can be a real nightmare. Neural networks can be used as an intermediary representation for solving a wide range of programming problems. By setting the appropriate weights a search space may be specified over some cost function with the solution at its minima. Invocation of the activation rule will lead to possible solutions. This basic form of (fine-grained) programming was pioneered by Hopfield and Tank (1986). Moreover, many tasks can be solved by learning from examples. Merely presenting a learning network with a representative portion of the task may cause it not only to learn this specific portion but also to generalize its behaviour to the entire task domain. Our research and that of others (see Part III) indicates that for the desired generalization behaviour to take place it is important that network topology be constrained, for instance by translating known aspects of the task into a broad – modular – network topology (coarse-grained programming; see also Hudson *et al.*, 1990). Training such a globally configured network will induce the desired fine-grained behaviour, with a high level of generalization. Such a learning approach might be called 'autonomous programming', because the role of the programmer can often be reduced to a minimum, even with complicated problems.

B1.4 A classification for hardware implementations

Many different implementation approaches have been tried in the past 5 years. In order to make the discussion in the following appendices more manageable, a broad classification of neural network implementations will be presented that may serve as a frame of reference. A recent and more extensive survey covering most of the literature on hardware implementations has been carried out by Treleaven (1989), and another good introduction is the somewhat 'older' DARPA report (1988). Table B1.1 gives a summary of the classification. As will be clear, many of the dimensions distinguished are interrelated. The scheme could be refined by making some of the hierarchical relations between the various dimensions more explicit; for instance, serial processing implies a virtual implementation. The nine dimensions distinguished are, of course, not

Table B1.1 Various dimensions for classifying neural network implementations.

Dimension	Distinctions		
General approach	Virtual	Hybrid	Physical
Technology	Optical	Electronic	Molecular
Machine architecture	Serial	Pipelined	Parallel
Temporal evolution	Synchronous	Asynchronous	Irregular
Precision	Binary	Discrete	Continuous
Control structure	Neural	Pseudo-neural	Algorithmic
Applicability	General purpose	Dedicated	
Learning mode	Fixed	Modifiable	*In situ*
Program control	MIMD	SIMD	

exhaustive, but they suffice for the purpose of organizing the following appendices. We will briefly discuss each of the dimensions.

In the overall organization here, a principal distinction is made between *virtual*, *hybrid* and *physical* approaches to implementation. This forms the first dimension of our classification. In a virtual implementation both nodes and connections are multiplexed over a relatively small number of processors. A technique often used is to assign a fixed number of nodes to a processor. Activations are exchanged between the processors through the communication channels provided by the machine. Thus, connections are multiplexed over available communication channels. In most cases the weights of the in-connections to a node are stored at the processor implementing that node. Of course, if only one processor is available (i.e. on a serial machine), all nodes and connections are multiplexed over this single processor. In hybrid implementations each node is implemented by a physical processor, but weights are multiplexed, for instance over a single communication bus or through a more complicated communication structure. In physical implementations both nodes and connections are mapped onto distinct hardware components. Most physical implementations (partially) rely on analog computation, although this is not a requirement.

A second dimension derives from the techniques that have been used to implement neural networks. These are 'classical' *electronics*, *optics* and a combination of both techniques: *electro-optical* implementations. In the future, *molecular* computing may be applied for the implementation of neural networks. Electronics comprises both conventional and experimental single and multiprocessor designs, as well as analog implementations (e.g. Treleaven, 1989). Purely optical designs are still rare. A few successful implementations have been described that use holograms to store weights (Abu-Mostafa and Psaltis, 1987). Pictures can be retrieved by presenting incomplete or distorted variants. Most optical implementations are actually electro-optical. One reason for this is that for most applications patterns must be transduced from a given format into some form of picture. For the general case, until general purpose optical computers are available, such transductions can only be accomplished in electro-optical systems. Another reason is that interleaving optical processing with electronic circuitry introduces the possibility of manipulating the signals in ways that still present difficulties in pure optics, for instance the implementation of a specific activation rule or the amplification of signals. Molecular computers have not yet been realized, although many techniques for doing so have been proposed (e.g. Conrad, 1985). Molecular computing is akin to neural networks in that it is also loosely modelled on certain principles found in biological reality, namely the interactions between molecules (although some approaches actually consider subatomic levels of interaction). The approach holds great promises but may take more than 15 years to develop (Treleaven, 1989).

A third dimension is between *serial* and *parallel* modes of processing. As a case in between we might refine the parallel mode into a fully parallel and a *pipelined* mode of processing. Among the serial implementations we also include the commercially available coprocessor boards.

A fourth dimension concerns the intentions with which a machine is built. Many (experimental) machines have been built with specific applications in mind. These are called *dedicated* in contrast to *general purpose* machines. An example of a dedicated machine is a multiprocessor called GRAPE, which was built by Sugimoto *et al.* (1990) for the solution of large instances of the N-body problem. The distinction pertains more to the intention of the developers than to the actual use. Some dedicated machines seem to be amenable to often unexpected applications. The experimental GF11 multiprocessor, for instance, was originally built at IBM for the solution of certain problems in quantum chromodynamics, but seems to be well suited for the implementation of neural networks (Witbrock and Zagha, 1989). In this work, if a certain machine is suited for the implementation of a wide range of neural network models it will be called *general purpose*, while in all other cases it will be called *dedicated*. The Connection Machine (Hillis, 1985, 1987) is an example of a general purpose parallel machine, and so are all conventional serial computers. The general purpose classification introduced here should not be confused with the idea of a general neurocomputer (see Section B5.6). The latter concept refers to a hypothetical machine on which a large range of problems can be solved by using a synthesis of neural network theory, parallel implementation technology and interface design.

A fifth dimension concerns program control on a parallel machine. This can be of the multiple-instruction multiple-data (*MIMD*) type, or the single-instruction multiple-data (SIMD) type (Flynn, 1972). Vector processors are examples of an SIMD program control: in principle, each element of the vector receives exactly the same treatment. SIMD machines are generally less suited for neural networks, although some very fast implementations have been realized (e.g. Singer, 1990). Multiprocessor machines, such as transputer networks (see Appendix B2), allow each processor to be programmed separately. These machines are, therefore, said to be of the MIMD type. MIMD machines offer better support for a wide range of neural network models, because they impose fewer restrictions on the processing order.

A sixth dimension we distinguish is the control structure of the neural network: *local*, *semi-local* or *algorithmic*. The main criterion for this classification is the extent to which a model adheres to the principle of locality: all information needed at a node (or weight or other object) must be conveyed through explicit connections. It will be clear that in neurophysiologically plausible models no 'magic' influences can be allowed. Only locally available information can affect processes. This information need not be conveyed through synapses, and many global influences are found. The important issue is that in the 'wetware' of the brain no opportunities exist for 'off-site' calculations. Examples of violations of the principle of locality can be found in many of the currently popular neural network models. For instance, the activation rule used by competitive learning (Rumelhart and Zipser, 1985) is non-local, because each node 'knows' whether it receives the highest total input (compared with the other nodes) or not. A similar non-local principle is also used in ART 1 (Carpenter and Grossberg, 1987, 1988; Grossberg, 1976) and in Kohonen maps (Kohonen, 1989a, 1990). In these particular cases the activation rule could be converted into a neurophysiologically more plausible rule by using continuous, decaying activations and lateral inhibition. Many learning rules are also dependent upon non-local information. An important example is the backpropagation algorithm (Werbos, 1974, 1988; Rumelhart *et al.*, 1986) in which the error propagation against the flow of activations through the (directed) connections forms a violation of the principle of locality. Another example is the use of non-local activations (i.e. from nodes that have no explicit connections) to normalize the connection weights (see also the discussion in Chapter 2 about the learning rule used in ART 1, and the local variant used in CALM). Numerous other examples of non-localities in particular models could be listed, such as the use of 'spread', where the weights at the same relative position in different receptive fields are kept equal during learning (e.g. LeCun *et al.*, 1990a). In most of these models it seems possible to find a more plausible variant, usually at the expense of mathematical tractability. Therefore, all of these models are classified as 'semi-local'. All models that have no clear relation with neural networks will be called 'algorithmic', even though in many cases they can explain many of the same phenomena (e.g. Wolpert, 1990a).

A seventh dimension is the way temporal control is implemented. Some models use an *asynchronous* activation rule, while most others use a *synchronous* rule. An example of a model that uses an asynchronous activation rule is discussed in Hopfield (1982). At every iteration *only one* randomly selected node is updated. It is of no use implementing such a network in parallel hardware, because all nodes have to be updated serially. For Hopfield networks some small modifications can be made that allow them to function synchronously (Ferscha, 1990); all nodes are updated at every iteration. Still other models use rather complicated schemes of sequential updates of network objects such as nodes and connections (e.g. backpropagation). These could be called *irregular*. In Appendix B5 some problems with the implementation of irregular temporal control structures are described. In neurosimulators running on general purpose machines it usually suffices to make explicit the inherent parallelism in these temporally irregular models. In dedicated implementations the temporal control structure of irregular models must be integrated with the general design.

An eighth dimension concerns the precision of coding information. Models can be classified on the basis of resolution required for the representation of activations and weights: *binary*, *discrete* and *continuous*. It is often possible to make binary or discrete approximations to continuous models (see Appendix B3).

Finally, as a ninth dimension, an important classification of hardware implementations is the mode of learning. Some hardware implementations use *fixed* weights (i.e. one-time programmable weights, such as in holograms). The most prevalent form of learning uses *modifiable* weights with 'off-site' calculation of the learning rule in a host computer. These weights may be either programmable (on RAM), or only reprogrammable (on EPROM). The third and most interesting mode is *in situ* or *self-adaptive* learning, in which the weights can be modified continuously, without necessitating 'off-site' calculations that interrupt and seriously slow down processing.

In the following appendices, three examples of CALM implementations in parallel hardware will be presented. In Appendix B2, a virtual implementation of CALM and of other neural networks on a transputer network will be analysed. Appendix B3 describes a hybrid implementation of CALM on a 400-processor machine, called the BSP400 (this machine has a modular architecture). In Appendix B4, some remarks on a possible analog implementation of CALM are made. Finally, in Appendix B5 three environments for the development of CALM and other neural networks are discussed: the CALM Development System, the CALMLIB and the MetaNet Network Environment. In that appendix some ideas for a general, parallel neurosimulator will also be outlined.

Appendix B2

Virtual Implementations on Transputer Networks

B2.1 Background on transputer implementations

A large number of transputer implementations of neural networks have been published in the past few years (e.g. Aiken *et al.*, 1990; Bakkers, 1988; Beynon, 1988; Ernst *et al.*, 1990; Ferscha, 1990; Fukuda *et al.*, 1990; Johannet *et al.*, 1987; Koikkalainen and Oja, 1988). In this appendix (see also Murre, 1991b,c), an implementation of the CALM Development System (see Section B5.2) on a T800 transputer network will be discussed, followed by an analysis of the performance of transputer implementations, focusing on the achievable speed-up with virtual implementations and on the optimal size of a processor network. In the implementation (Van den Bout, 1990), nodes were multiplexed over the available transputers (see also Section B2.2). Routing functions were developed for the distribution of data (i.e. activations) to all processors at each iteration. The general analysis in Section B2.2 shows that for modular neural networks such as CALM, more efficient implementations can be developed. This possibility has not yet been pursued, but remains one of the objectives of a future project (Hudson and Murre, 1991). First, a brief description of the T800 transputer will be given, followed by some remarks on the implementation of the CALM Development System. The remainder of the appendix presents a general analysis of neural network implementations. As will be shown, such an analysis can be useful not only for estimating optimal implementation parameters, but it may also contribute towards a better understanding of implementation constraints in the brain, in particular those concerning connectivity patterns.

The T800 transputer is a multiprocessor building block produced by the INMOS division of SGS-Thomson. With a little support circuitry, networks consisting of several T800s can easily be constructed. Such networks are usually connected to a host computer. A transputer network is a MIMD machine, because each processor is separably programmable. The T800 processor has 4 kbytes of fast on-chip static RAM, a 32 bit integer central processing unit (CPU), and a 32 bit IEEE floating-point unit (FPU). The CPU and FPU can work fully in parallel. Furthermore, the T800 has four link interfaces. These links are direct-memory-access-controlled, bidirectional, serial-transmission channels which can provide up to 9 Mbytes of I/O throughput for a transputer. Transputers can be configured in a variety of topologies, provided that the fan-in/fan-out to each processor is equal to or less than four (i.e. the number of links). This allows for a four- (but not five-) dimensional hypercube, a binary or ternary tree, a flat grid or torus (see Figure B2.1), and many other designs. These configurations are usually hardwired, although some systems are available that allow modification of the topology using programmable link crossover switches.

The 'native language' of the transputer is Occam. The Occam language is based on Hoare's communicating sequential processes and allows succinct expression of parallellism in an algorithm. The transputer is designed to execute concurrent 'Occam processes', i.e. multiple threads of execution which one transputer can execute by timeslicing, assisted by special hardware. This feature can be exploited to optimize interprocessor communication: some processes may work on the data while others

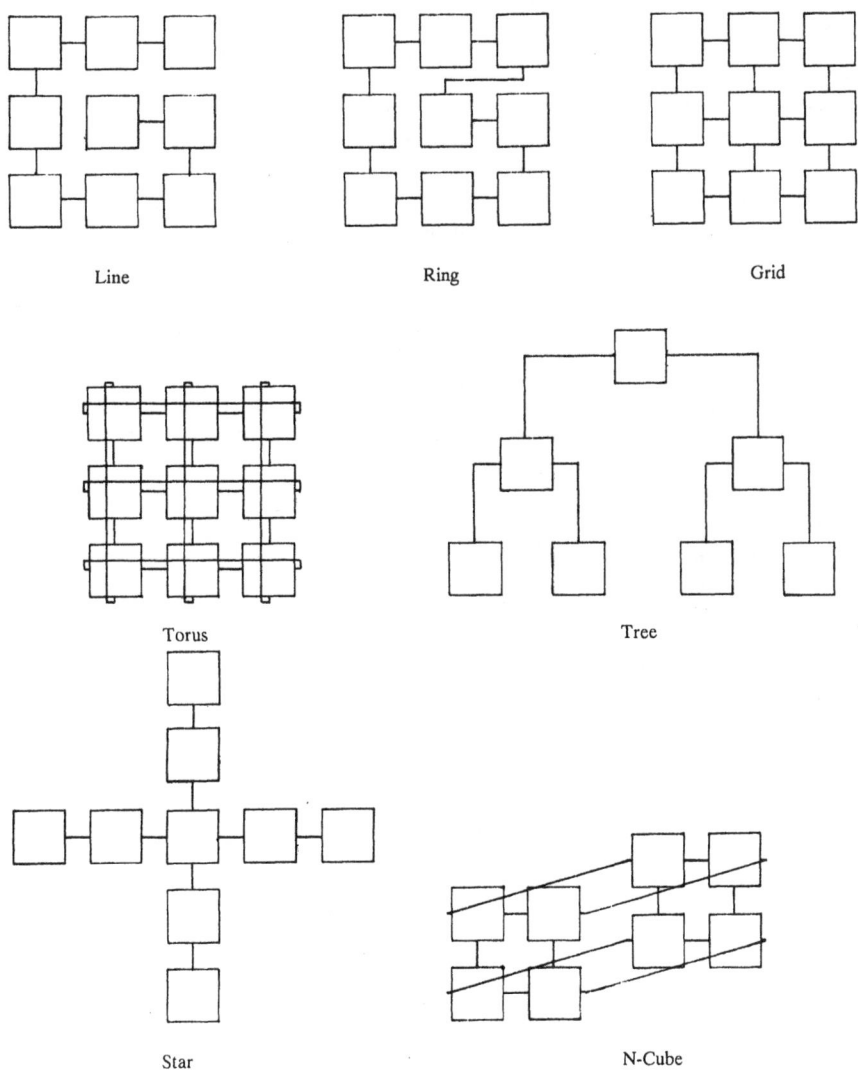

Figure B2.1 Different processor topologies.

communicate with neighbouring transputers. Occam supports parallel processes by providing special keywords such as PAR and SEQ. Parallel processes are initiated by definitions similar to the following:

```
PAR
    SEQ
        do process 1
    SEQ
        do process 2
```

These statements will start two parallel processes. Parallel 'Occam processes' communicate through Occam *channels*. Inputs and outputs can occur simultaneously over the interprocessor links. Processes on neighbouring transputers can communicate through so-called *hard channels*, which are mapped to the communication links provided by the hardware.

Several dialects of existing computer languages such as C and Fortran have been developed for the transputer. In our implementation of CALM we have used Par.C, a transputer dialect of C developed by ParSec in Leiden. The major advantage of using dialects like Par.C is that only a few new statements have to be learned (given knowledge of C), not an entire language. Furthermore, it is usually fairly easy to single out critical sections in existing, large conventional programs. Parallelizing these sections only may greatly improve overall performance. In this way, the 'dusty decks' of old software can be converted into reasonably optimal parallel code. Par.C adheres to most of the ANSI C standards (Kernighan and Ritchie, 1988). In addition to the regular C keywords, a few extra have been added to take care of transputer-specific operations. These keywords function much like similar Occam keywords (Dowsing, 1988). Typical Par.C keywords are: *par*, *channel* and *select*. The Par.C syntax corresponding to the Occam example given above would be:

```
par
{
    Process1();
    Process2();
}
```

The two processes are executed simultaneously (on one transputer). The *channel* type works in a similar way as in Occam. It could, for instance, be used to communicate between *Process1()* and *Process2()* in the above example. Other keywords (*select*, *alt*, *cond*, *guard* and *timeout*) add more flexibility for interprocess communication.

For communication between processes running on different transputers Par.C provides only two basic functions: *RecvLink()* and *SendLink()*. These functions enable communication over any of the four links on a transputer. They do not provide for direct interprocess communication. As an example, the following statement receives an array of integers called *Data* over link 2:

```
int Data[400];

RecvLink(2,Data,400*sizeof(int));
```

As can be observed, it is necessary to specify in advance the number of bytes being received. In many cases, this is a very impractical situation. Either the same amount of data must always be sent, or separate messages (of known size!) must announce the size of the following incoming block of data.

Additional difficulties with interprocessor communication are caused by the fact that, if two processes on neighbouring transputers A and B must communicate, it is necessary to synchronize these communications explicitly. The reason for this is that several other processes may be operating on both transputers, each of which may want to transfer data over a certain link. At the time of our implementation project (early 1990), Par.C did not offer any additional support for communication between processes at two transputers that are not directly connected. Unfortunately, a large part of our implementation project had to be devoted to the development of a set of data-routing functions (Van den Bout, 1990). The goal was to develop a number of functions that would enable easy interprocess communication on different transputers which were possibly not directly connected, irrespective of the topology of the transputer network.

The main problems with routing were finding the shortest path from transputers A to B through the network and synchronization of communications (Van den Bout, 1990). The standard algorithm for finding the shortest path is based on the following algorithm (Dowsing, 1988). Suppose that each transputer has a list that specifies which link of the four links to take in order to reach a certain goal transputer through the shortest path. A block of data can then easily be sent along the shortest path by choosing the appropriate link at each intermediate transputer. The compilation of such a directions list for each transputer can proceed in parallel. Initially, on a certain transputer the distance to all other transputers on the directions list is set to infinity (and to itself to zero). At each of the following steps, all transputers exchange their lists. On the basis of the (maximally) four incoming lists a transputer is able to update its list by finding all the entries which have a distance of less than infinity. A distance entry d in a newly received list is incremented (since it traversed one extra link) and copied to the directions list only if $d + 1$ is less than the value of the current entry in the list. For example, if in the directions list on transputer 17 it says that the distance to 23 is infinite and a list is received through link 3 that has a distance of 4 to transputer 23, then in the directions list of transputer 17 it will be recorded that the shortest path to 23 is $4 + 1 = 5$, starting at link 3. The list exchange must be continued for k steps, where k is the maximum intertransputer distance. This distance is usually not known, so that k is taken to be n, the total number of transputers.

In order to ensure that transmitted data have reached their destination, it is necessary to synchronize communications. For this purpose a number of synchronization routines were developed. They make it possible for a transputer A to check whether a certain transputer B has indeed received all of the data sent by A. This is done by requesting a confirmation after a certain number of data transmissions. In addition to synchronizing communication, congestion deadlocks must be prevented. There is a danger of deadlocks when all buffers for receiving data are full, preventing communication between any two transputers, because neither can receive data. In the development of the data-routing functions special precautions had to be taken to prevent such situations. This was accomplished by using a cycle through all transputers in the network (possibly in both directions through a certain link). As long as buffer space is available somewhere in the cycle the total congestion of communications can always be prevented.

Only after the development of all basic communication software was a neural network implementation feasible. The CALM Development System (see Section B5.2) was used as the 'dusty deck' to be parallelized. The parallelization technique used was multiplexing nodes over processors, i.e. the array that held the data of the m node structures (activations, incoming weights, attributes, etc.) was divided into n partitions, each of which was assigned to a processor. After each iteration the new activation values were exchanged between processors and calculation processes were synchronized. The exchange of weights was not necessary, because the weights remain local at the nodes.

Program development proceeded much slower than desired due to problems caused by bugs. A major problem with debugging programs on the transputer is that it is virtually impossible to pinpoint which process went wrong and at what time. The reason for this is that many processes may be running simultaneously on any transputer. Moreover, only one transputer can send output to the host computer. Sending output will itself start a new process, possibly involving communication over several other transputers in order to reach the host. This process may interfere with existing error conditions, so that bugs disappear as soon as one tries to detect them (but only to reappear somewhere else). A major conclusion of the small implementation project is, therefore, that it is absolutely crucial that debugging tools be developed for transputers and similar systems. In the remainder of this appendix, an analysis of transputer implementations of neural networks will be presented.

B2.2 Performance analysis

In the analysis below we will assume that the m nodes of a neural network are multiplexed over n processors, where n is less than m. Each processor handles one partition of nodes. All incoming weights to a node are stored at that node only. Following each iteration, the activation of a node A must be sent to all partitions that hold nodes with incoming connections for A. If a neural network is distributed over the processor network in this manner, in the absence of any communication overhead and assuming efficient load balancing, the speed-up would be equal to the number of processors, n. If we let c be the time to complete one iteration with the entire neural network running on a single processor (i.e. the case where $n = 1$), then, in this ideal case with n processors, the iteration time would decrease to c/n.

Completion of one iteration may involve the execution of several communication processes on each transputer. The contribution of communication overhead turns out to be a major limiting factor in the efficiency of implementations on transputers and similar MIMD machines. It appears that with fully connected neural networks little can be done to alleviate this problem.

With fully connected neural networks, the effect of topology on data-transfer time is small. Whereas it is generally true that the topology of a processor network (i.e. grid, line, hypercube, etc.) strongly influences the communication overhead of a certain parallel program, it can be shown that for the implementation of fully connected neural networks (e.g. Hopfield nets) this assumption does not hold. At best, there may be a small, constant reduction. The main reason for this is that, for such networks, all partitions of nodes (i.e. the activations) *must be sent to all other processors.*

Let us assume that all partitions contain the same amount of data, and that sending and receiving one partition over a link between two processors takes one unit time step. (A transputer may send and receive in parallel.) Note that the minimal transfer time in the processor network is bounded by the longest transfer path. Let us first study the worst-case topology.

Obviously, a *line* topology has the longest path (also called the diameter of the processor network), so we will consider it to be the worst topology. Call the two outer-most processors A and Z. If we assume that at each time step all processors swap one partition over both their links, then, after exactly $n - 1$ unit steps (i.e. swaps), the partition from processor A will have arrived at Z, and vice versa. At that moment, all other partitions will have been distributed to all other processors. Thus, for the worst topology, transfer takes $n - 1$ unit steps.

Now, assume that some more efficient processor topology exists. Take any processor in this hypothetical network. No matter what the characteristics of the topology are, at every time step the selected processor can only receive and send four partitions (as there are only four links on a transputer). With n processors in the network, it will therefore take $(n - 1)/4$ unit steps to transfer all data in the processor network. The conclusion must be that with a fully connected neural network, optimizing the topology of a transputer network can at best reduce the data-transfer time with a constant factor of four. This result applies to any multicomputer network where the processors are con-nected by a fixed number of links over which they can, at each time step, exchange a constant amount of information (as opposed to, for instance, a bus-oriented system). If we call the number of links g, then, with the optimal topology for such systems, data-transfer time will never be more than g times faster than with a line topology.

It should be pointed out that reduction of the diameter (longest path) of a topology does not guarantee even a constant gain. For instance, if we were to modify an n-processor line topology to an n-processor star topology, we would still need $n - 1$ unit steps, as the reader can easily verify. For processor types that have serial redistribution of data over multiple links, choosing the wrong topology of the processor network may actually *increase* communication time by a factor g, where g is the number of links minus one. For instance, if we compare a four-star topology consisting of nine proces-sors with a ring topology of equal size, then the central processor in the star must send (redistribute) 24 partitions, whereas in the ring any processor only redistributes eight partitions. The star topology is less effective because the central processor forms a com-munication bottleneck. On an N-cube, torus and ring redistribution is optimal, because these topologies have no bottlenecks. So, we cannot assume that topologies with a lower diameter or higher connectivity necessarily reduce communication overhead, at least not in the general case.

Randomly connected neural networks and fully connected neural networks have comparable communication overhead. We will now show that the above remarks on the effect of topology also hold true for (sufficiently large and dense) randomly con-nected neural networks, even if the nodes have fixed fan-in, where the number of incoming connections to a node is small and constant. This is a somewhat surprising result because one would expect that fewer data need to be transferred in networks with fixed fan-in. Even with strong limits on connectivity, however, the attainable reduction of communication overhead is negligible. To prove this, it suffices to show that in these networks a partition P_A at a processor A must still be sent in full to all other processors.

Consider the case where we have two partitions P_A and P_B each with z nodes. Sup-pose that each node in P_B has a constant partial fan-in of exactly f_p nodes in P_A. That is, each node in P_B has f_p incoming connections from nodes in P_A. At the end of an iteration, activations of nodes in P_A that have outgoing connections to *any* node in P_B must be transmitted. Or, similarly, an activation of a node in P_A does not have to be transmitted if no node in P_B receives any connections from it.

If we assume random interconnections between P_A and P_B, we can estimate the chance that a node in P_A has no outgoing connections to (any node in) P_B if z is large.

Intuitively, we could argue that, with large z, the chance that a processor is not included in the fan-in bundle to a node in P_B becomes larger, because the number of nodes in P_A increases. Alternatively, we could argue that, because more fan-in bundles are sent from P_A to P_B, this chance becomes very small. As it appears, the answer is somewhere in between, although it approaches the latter option.

The chance that a given node in P_B receives a connection from a given node in P_A is equal to the proportion of the fan-in relative to the size of the partition, z. Thus, the chance of no connection is

$$1 - \frac{f_{\mathrm{p}}}{z} \tag{B2.1}$$

Because the fan-in connections to P_B are independent, the chance that a given node in P_A has no connections to any node in P_B is

$$\left(1 - \frac{f_{\mathrm{p}}}{z}\right)^z \tag{B2.2}$$

Now, if we let the partitions become very large we obtain

$$\lim_{z \to \infty} \left(1 - \frac{f_{\mathrm{p}}}{z}\right)^z = e^{-f_{\mathrm{p}}} \tag{B2.3}$$

The conclusion must be that, if partitions and fan-in are sufficiently large (e.g. $z > 50$, $f_{\mathrm{p}} > 5$), there is only a small chance that some activation need not be transmitted from a given partition to another. This is the same as saying that almost all activations must be transmitted from a given partition P_A to any other partition P_B.

Thus, even with fixed fan-in in randomly connected neural networks, all partitions must be sent to all other partitions. Therefore, the remark about the slight influence of processor topology on data transfer applies equally to these networks.

B2.3 Analysis of fully or randomly connected networks

The total time spent on communication events on one transputer may be decomposed into a constant portion t_k and a variable portion t_v. On a transputer, initiating a communication process takes a certain, constant amount of time k. This is one of the parameters in t_k. The time taken to send a block of data, which is proportional to its size (i.e. in bytes), is a parameter in t_v. If we call the iteration time on one transputer t_i, we obtain the following general expression

$$t_i = t_k + t_v + \frac{c}{n} \tag{B2.4}$$

In order to estimate the variable and constant portions of the communication time we must consider the algorithm used to route data through the network.

The above results on the effects of topology allow us to use a straightforward routing algorithm, in which data in a certain partition is sent to all others. It should be pointed out that if we use such an algorithm we must restrict our analysis to fully or randomly connected neural networks. For neural networks with regularly constrained architectures, more efficient routing schemes can be devised.

As described above, we have used a routing scheme where a full cycle through the network is constructed, possibly in both directions of the bidirectional links. Thus, the network can be viewed as a tree with one processor as the root. The circle is started at the root, curves around the branches of the tree, and returns to the root. This scheme ensures that no transfer deadlocks occur as long as at least one buffer in the circle remains empty.

As the above argument demonstrates, any processor will be engaged in $n - 1$ communication processes. At each step any two directly connected processors may swap two partitions over their link. Of course, a partition is only sent once over a link. From now on, we shall ignore the constant speed-up of maximally a factor of four to be gained with certain topologies. The effects may be included in the constants of the equations below. This brings the constant portion of the communication time to

$$t_k = k(n - 1) \tag{B2.5}$$

where k is the constant time to initiate (and end) a single communication process.

The total number of activations sent through the network is $m(n - 1)$, where m is the number of nodes in the neural network, because all activations on a processor must be sent to all $n - 1$ processors (but not to itself). The variable portion of the communication time is thus

$$t_v = \frac{rdm(n - 1)}{n} = rdm\left(1 - \frac{1}{n}\right) \tag{B2.6}$$

where r is the precision of the representation in bytes of a single activation value (i.e. when a double-precision representation is used $r = 8$ bytes), and d is the time taken to send 1 byte over a link (we measured $d = 1 \mu s$). If we ignore the terms due to the exclusion of self-communication of data on processors from the communication times t_k and t_v (i.e. in equations (B2.5) and (B2.6), for large n, $n - 1$ becomes n) we arrive at the following composite iteration time:

$$t_i = kn + rdm + \frac{c}{n} \tag{B2.7}$$

where all constants and variables are as above (c is the iteration time on one processor).

The optimum number of processors $\tilde{n}$ can be found by taking the first derivative of equation (B2.7) with respect to n and setting it equal to zero. We then obtain

$$\tilde{n} = \sqrt{\frac{c}{k}} \tag{B2.8}$$

To simplify the analyses, we propose that the iteration time on a single processor depends primarily on the number of weights to be processed. For most types of neural networks this is not an unreasonable assumption. The processing of the weights usually involves the calculation of the sum of the weighted input activations. For learning neural networks, it also involves the application of a learning rule to each individual weight. In the absence of specialized neural network hardware (e.g. large vector processors), the total time spent on processing the weights will, therefore, be linearly dependent on the number of weights. The processing of the activation (or any associated error term or delta) is always limited to the application of an activation rule, which takes less than some given constant amount of time. Moreover, this amount of time will be small. It will usually involve a few multiplications, and perhaps occasional division operations. Because the processing of a single weight will also involve at least one multiplication operation, the time needed for processing a single activation will be equal to the time needed to process a small number of weights. The precise number is, of course, dependent on the type of activation and transfer rules used, and on whether a learning rule is applied.

Now, if we let the number of weights in the network be Fm, where F is the fan-in per node (i.e. the number of incoming connections), in fully connected networks $F = m$. In network types with a constant fan-in, F is equal to some constant number f. With constant fan-in we thus have

$$c = afm \qquad (B2.9)$$

where a is the time needed to process one connection between any two of the m nodes. In the processing time for one connection, we include the calculation of its contribution to the total input activation (i.e. multiplying a weight with the incoming activation value) and application of a learning rule. If we substitute equation (B2.9) in equation (B2.8) we obtain

$$\tilde{n} = \sqrt{afm/k} = \beta\sqrt{fm} \qquad (B2.10)$$

with

$$\beta = \sqrt{a/k} \qquad (B2.11)$$

Equation (B2.11) shows that, if the constant portion of the communication time k increases, the number of transputers that can be effectively used will decrease. In general, $\tilde{n}$ increases with m, the number of nodes in the network. But only in fully connected networks, with $F = m$, do we have a linear relation

$$\tilde{n} = \beta m \qquad (B2.12)$$

The value a is highly dependent on the specific type of learning and activation rules used. We estimated it to lie somewhere between 10 and 250 μs. This corresponds to an approximate processor speed of 4000 to 100 000 connections per second (CPS). The exact value depends on the type of neural network. On our transputer network we measured 250 μs for k, so β will lie between 0.2 and 1.0. With these figures, a fully connected neural network with, say, 1000 nodes will function optimally on 200 to 1000 transputers, depending on the time it takes to process a single weight. With a limited fan-in of, say, $f = 80$, the optimal size of the transputer network will range from 57 to 289 transputers. Note that the optimal size is defined exclusively in terms of minimization of the iteration time; it does not include considerations such as the price of the system. Indeed, the last few added transputers will make only a small contribution towards a reduction of the total iteration time.

The maximum speed-up achievable on a fixed-size transputer network can be found by considering the ratio between the iteration time on a single processor, c, and the iteration time of the entire processor network t_i (i.e. c/t_i), and letting m become very large. Let us first consider the case where we have a (variable) fan-in F, the size of which depends on the number of nodes: $F = m^b$, with $0 < b \leqslant 1$. The ratio then becomes

$$\frac{c}{t_i} = \frac{am^{b+1}}{kn + rdm + (am^{b+1}/n)} \qquad (B2.13)$$

and with m very large we obtain

$$\lim_{m \to \infty} \frac{nam^{b+1}}{kn^2 + rdmn + am^{b+1}} = n \qquad (B2.14)$$

In large, randomly or fully connected networks (i.e. $F = m$) the achievable speed-up is proportional to the size of the transputer network. In the case of fixed, limited fan-in with $F = f$ and $f > 0$, the limit becomes

$$\lim_{m \to \infty} \frac{nafm}{kn^2 + rdmn + afm} = \frac{n}{1 + n(rd/af)} \tag{B2.15}$$

So, with fixed fan-in, if n becomes large (but with $n \ll m$), speed-up will approach the constant fraction af/rd. For example, if we take $a = 100 \ \mu s$ (i.e. a processor speed of 10 000 CPS), fixed fan-in $f = 80$, double precision $r = 8$ (i.e. 8 byte representations), time to transmit 1 byte $d = 1 \ \mu s$ (as measured by us), then the ratio af/rd would equal 1000. In a large transputer network of 1000 processors, by substituting all of these figures in equation (B2.15) we obtain a speed-up of $1000/(1 + 1) = 500$. For small processor networks (i.e. small n) the speed-up scales nearly linearly, because the fraction nrd/af remains small, as in the case where $F = m$ (equation (B2.14)).

B2.4 Analysis of modular networks on a torus

Whereas in neural networks with random connections there seems to be no straight-forward way to benefit from the restriction of limited fan-in, some types of neural network offer the possibility of reducing communication overhead. We will consider here the case where each partition has (direct or indirect) outgoing connections to only a limited number of other partitions. This condition is satisfied by many modular neural networks. A central theme in this book is that modular networks are preferable from a biological and psychological point of view. The results below offer additional support for the modularity principle from an implementation point of view.

In our definition of a modular network we assume that a balanced distribution of modules over processors is possible, and that partition fan-out is a constant number s for all processors. Thus, all processors send their partition to s randomly distributed goal processors. We assume, furthermore, that s is small compared with n. To simplify the calculations we finally assume that partition fan-in is also equal to s. These assumptions state that the inherent modularity in the neural network can be used to arrive at a more optimal implementation. The analyses below are valid only to the extent that this proposed optimization is actually achieved. If modules in the network contain only a few nodes, and if the modules are densely interconnected, it is unlikely that the implementation can benefit much from the (slightly) more regular structure of the network.

For modular neural networks in the above sense, however, the topology of the processor network can strongly reduce data-transfer time. The most relevant topology for a transputer network to consider here is a torus (see Figure B2.1), because it is highly regular, densely interconnected, uses all four links available at any transputer, and because it is most widely used for transputer networks. On a torus, the maximum path-length is $\sqrt{n}$ processors. Suppose that we use a routing algorithm where at each transputer a check is made on the partitions simultaneously available at its four links, so that data that have to be relayed to a goal transputer at maximum distance are handled first. This scheme will ensure that the total iteration time is equal to, or less than, the time taken to transmit the partition that has to travel the longest distance. As for the ring topology, on a torus for large n and s, the longest distance will be close to the maximum distance. If we take the longest distance as an upper bound, then a worst-case estimate for the total pathlength travelled by a given partition on a certain processor to all its goal processors is $s\sqrt{n}$. The total amount of data transmitted (and received) in the network by all processors will, thus, be given by

$$nrs\,|\,P\,|\,\sqrt{n} \tag{B2.16}$$

With $|P| = m/n$ as the size of a partition, the average amount of data to be transferred by each processor is

$$rms/\sqrt{n} \tag{B2.17}$$

Equation (B2.17) gives an upper bound and ignores possible savings due to common paths. If $s \ll n$, this appears to be a reasonable assumption. If we use a routing scheme of the type outlined above, the number of processes that have to be initiated by each processor is at most $s\sqrt{n}$, so that for the total iteration time we have

$$t_i = ksn^{1/2} + rdsmn^{-1/2} + cn^{-1} \tag{B2.18}$$

where k, d and c are as in the analysis above. The optimum number of processors $\tilde{n}$ by a fixed, large m can be found by taking the first derivative with respect to n and setting it equal to zero:

$$\tfrac{1}{2}ksn^{-1/2} - \tfrac{1}{2}rdsmn^{-3/2} - cn^{-2} = 0 \tag{B2.19}$$

If we multiply equation (B2.19) by $2n^2/ks$ we obtain

$$n^{3/2} - \frac{rdmn^{1/2}}{k} - \frac{2c}{ks} = 0 \tag{B2.20}$$

Now, if we set $p = -rdm/k$ and $q = -2c/ks$, and $z = \sqrt{n}$, equation (B2.20) reduces to

$$z^3 + pz + q = 0 \tag{B2.21}$$

With $c = aFm$, where F is the (variable) connection fan-in per node, using Cardan's formula, we find

$$\tilde{n} = \left\{ \left[\frac{aFm}{ks} + \left(\frac{a^2F^2m^2}{k^2s^2} - \frac{r^3d^3m^3}{27k^3} \right)^{1/2} \right]^{1/3} \right.$$
$$\left. + \left[\frac{aFm}{ks} - \left(\frac{a^2F^2m^2}{k^2s^2} - \frac{r^3d^3m^3}{27k^3} \right)^{1/2} \right]^{1/3} \right\}^2 \tag{B2.22}$$

If we take $F^2 > m$, then for m sufficiently large we have

$$\frac{a^2F^2m^2}{k^2s^2} \gg \frac{r^3d^3m^3}{27k^3} \tag{B2.23}$$

so that the optimal number of processors becomes

$$\tilde{n} = \left(\frac{2aFm}{ks} \right)^{3/2} = \gamma(Fm)^{2/3} \tag{B2.24}$$

with $F^2 > m$ and

$$\gamma = \left(\frac{2a}{ks} \right)^{2/3} \tag{B2.25}$$

If we compare equation (B2.24) with the optimal number of processors in the case of fully connected neural networks, as given in equation (B2.8), $\tilde{n} = \beta\sqrt{(Fm)}$, we see that for large m more processors can be used effectively due to the reduced communication overhead in the modular case.

As above, we can derive the speed-up with large m by taking the following limit:

$$\lim_{m \to \infty} \frac{c}{ksn^{1/2} + rdsmn^{-1/2} + cn^{-1}} \tag{B2.26}$$

With large, node-dependent fan-in ($F = m^b$, with $0 < b \leqslant 1$), we have

$$\lim_{m \to \infty} \frac{am^{b+1}}{ksn^{1/2} + rdsmn^{-1/2} + am^{b+1}n^{-1}} = n \tag{B2.27}$$

With limited, constant fan-in, $F = f$, we derive

$$\lim_{m \to \infty} \frac{afm}{ksn^{1/2} + rdsmn^{-1/2} + afmn^{-1}} = \frac{n}{1 + (rds/af)n^{1/2}} \tag{B2.28}$$

(Note that a limited partition fan-in of s need not imply limited, constant node fan-in.)

As in randomly connected networks, we find that, if node fan-in is not kept constant, the speed-up due to added transputers is of order $O(n)$ for large neural networks. If node fan-in is limited, however, modular neural network implementations with limited partition fan-in have a markedly better speed-up of $O(\sqrt{n})$, compared with $O(1)$ for randomly connected networks (equation (B2.14)).

Table B2.1 summarizes the results of the performance analysis. With fixed node fan-in, performance seems to deteriorate. This apparent reduction in performance is caused by the fact that limiting fan-in decreases the processing load, whereas it has no effect on the communication load (see the discussion of randomly connected neural networks above), which is the major limiting factor on performance.

Although transputers scale well, the use of such a highly sophisticated (and relatively expensive) machine for the limited task of processing neural networks, in our opinion, may be a waste of resources. The main reason for using processors like the T800 for this task seems to be their availability and the relative lack of comparable multi-processor systems on the market. This stresses the necessity of developing neuro-computers especially for the task of processing neural networks. Only with dedicated neurocomputers may we hope to achieve the cost/performance ratio required to fuel the field of neural networks.

Table B2.1 Summary of the results for optimal size of the processor network and for speed-up with added transputers. In the table, n is the number of transputers, m is the number of nodes in the network, f is the size of the fixed fan-in, and F is the variable fan-in, with $\sqrt{m} < F \leqslant m$.

| | Connectivity | | | |
| | Full or random | | Modular (on torus) | |
Fan-in	Variable	Fixed	Variable	Fixed
Optimal size	$O(m)$	$O(\sqrt{fm})$	$O([Fm]^{2/3})$	–
Speed-up	$O(n)$	$O(1)$	$O(n)$	$O(\sqrt{n})$

B2.5 Concluding remarks

Some widely held intuitions about the importance of minimizing interprocessor distance and connection fan-in (to nodes) are inappropriate for fully and randomly connected neural networks. It was shown, however, that if modular neural networks are used, these factors can contribute significantly towards a reduction of iteration time, and a more effective use of processors. The major conclusions from the implementation project and analysis are as follows:

1. Software support for parallel processing on transputer networks, especially for debugging, was found to be inadequate.
2. Communication overhead is the decisive factor in the efficiency of scaled-up processor networks.
3. With fully or randomly connected neural networks, processor topologies that aim to reduce (processor) network diameter can only result in a small, constant reduction of communication overhead.
4. With randomly connected neural networks, limiting node fan-in does not generally lead to a decrease in communication overhead, even if fan-in is restricted to a constant number f, where f is a constant (say, greater than 2).
5. The global regularity in the connectivity pattern found with modular neural networks (as defined above) may be used to implement more efficient routing schemes.

Conclusions 2 to 5 are not restricted to transputer networks. In fact, they apply to any multicomputer implementation, in which processors can communicate only through a fixed number of links (as opposed to, for instance, a bus-oriented approach or the use of dedicated communication hardware for this particular task; see next appendix). The conclusions all bear on the necessity to execute a large number of point-to-point communications at each iteration of the neural network.

These results can also be translated into terms of analog hardware. For 'communication overhead' one must then read 'number of hardwired node-to-node connections'. We could suppose that, in a hardware implementation, nodes are distributed over a large number of processing clusters, and that these clusters have a limited number of interconnecting tracts (comparable with the four links of a transputer). With fully or randomly connected neural networks, any cluster has a total number of incoming wires equal to the number of nodes in the network (minus those already in the cluster). Different topologies of processing clusters, of course, do little or nothing to diminish the total size of the incoming tracts. As in the conclusions above, limiting node fan-in with randomly connected neural networks has a negligible effect on the total size of the tracts. This makes large hardware implementations of such networks unfeasible. For modular neural networks (as defined above), however, not all clusters need to be interconnected, so that the total size of the incoming tracts can be reduced.

The 'wetware' of the brain is subject to many of the same constraints as any hardware implementation. In particular, the space available for interconnections is limited. Nelson and Bower (1990) remark that if the 10^{11} neurons in the brain were put on the surface of a sphere, with full interconnections running through the sphere, its radius would have to measure over 10 km to accommodate the volume of connecting axons (of radius 0.1 μm). It is well known that neurons have a limited fan-in of about 10^4 incoming connections (Kandel and Schwartz, 1985).

From the above considerations, it follows that even though the neurons in the brain have limited fan-in, they cannot possibly have a random connectivity pattern.

Connectivity must be constrained in additional ways. At least two types of limited connectivity have been observed: modularity and locally dispersed connections. In Part I, in particular, it was argued that a lot of evidence supports the importance of modularity as an organizing principle in the brain, and that modular neural networks not only are most realistic from an implementation point of view, but also may be favoured from a psychological and biological perspective.

Another, more restrictive form of limited connectivity has only locally dispersed connections. In models with such a structure, nodes are located in two- or three-dimensional maps, where each node is connected to only a small number of nodes in the direct neighbourhood. Such structures give rise to computational brain maps (i.e. Kohonen, 1989a, 1990) which have been extensively documented in the literature (e.g. see references in Kohonen (1989a, 1990)). Nelson and Bower (1990) compare the functioning of brain maps with the often regular structures arising in processes on parallel computers. They argue that brain maps are computationally efficient for local computations, which are a common feature of many sensory processing tasks. And because they also minimize the size of interconnection tracts (or volume devoted to connections) they are likely to be selectively favoured. Not all types of tasks are well handled by computational maps, however, so that long-range connections remain necessary.

In short, arguing from an implementation point of view, randomly interconnected networks cannot be realistic models of the brain. We should, therefore, devote more attention to the study of models that in addition to local constraints, such as limited fan-in and locally dispersed connections, impose global limitations on connectivity, such as a modular structure.

It was argued above that transputers are a feasible, although expensive, approach to implementing neural networks, and that only dedicated neurocomputers will eventually be able to sustain research and application in neural networks. In Appendix B3, a realized study for the development of such a system will be described: the BSP400, a hybrid implementation of CALM. This system has a modular architecture to ensure scalability.

Appendix B3

Hybrid Implementation: The BSP400, A Dedicated Multiprocessor*

B3.1 The Brain Style Processor

This chapter describes an experimental, hybrid multiprocessor with 400 processors: the BSP400 (Brain Style Processor with 400 processors). As defined in Appendix B1, in a hybrid implementation each physical processor implements one node in a neural network. The node-to-node connections, however, are multiplexed over a limited number of hardwired communication channels. In the BSP400, processors are connected (only) by a single communication bus. This means that, at each iteration, all activations must be transmitted sequentially over the bus, according to a certain communication protocol (multiplexing).

The BSP400 was primarily developed for the implementation of the CALM model and similar types of neural networks. The BSP400 project was intended to be a feasibility study for a large-scale, general neurocomputer. The programmable microprocessors used serve to define specifications for VLSI, taking account of both local process and communication requirements. Rather than constructing a system with a few powerful processors we focused on a neurocomputer with many elements, each possessing limited processing capabilities.

The main design problem is scalability. A second problem is extendibility. A modular architecture may offer a possible solution to both of these problems. As shown in Appendix B2, the global regularity in the connectivity structure of modular neural networks can be used for more efficient virtual and physical implementations. Part of the BSP400 project aims at investigating for what architectures modularity in neural networks can be used to reduce communication bandwidth in hybrid systems. To enable extendibility, the system was designed as a number of units connected by a communication bus (see Figure B3.1). The system can be extended – up to a certain limit – by simply adding units. At the time of writing (January 1991) the BSP400 project is nearing completion. After its completion, it will be applied to a selected number of real-world problems (Hudson *et al.*, 1990).

---

* The research described in this chapter is a joint project of the Leiden Connectionist Group at the Unit of Experimental and Theoretical Psychology (Hudson, 1988; Hudson *et al.*, 1990) and the Department of Computer Architecture at the Technical University of Delft. Part of the work on the BSP400 has been described elsewhere, see for instance Heemskerk *et al.* (1991), Hoekstra (1990), Hoekstra *et al.* (1990); see also Murre *et al.* (1991).

As pointed out above, the BSP400 can be characterized as a hybrid multiprocessor (see also the classification in Section B1.4). It is a dedicated system in so far that with the current node software only CALM-type networks are implemented. It remains possible, however, to reprogram the processors' EPROM so that other networks can be run, for example using Ferscha's (1990) synchronous variant of Hopfield's (1982) model. Calculations and updates are synchronous. Activation and learning rules must be fully local; non-local control structures are not supported. The processors need not be programmed with the same node software. Activation values are transmitted as single-bit values, but the node–internal activation state is represented by 8 bits. Weights are both transmitted and represented as bytes. The approach taken here can, thus, be classified as a mixture of binary and discrete precision. The weights are kept in RAM memory and are updated during each processing cycle, so that the system is fully self-adaptive.

This appendix gives a brief introduction to the BSP400. First, the global architecture of the system will be described, followed by a more detailed presentation of the principal components and the way they cooperate in the calculation process. Then, a discrete version of CALM used in the implementation will be discussed, followed by a description of the user interface. The performance of the system will be analysed. The appendix concludes with some remarks on scalability, flexibility, extendibility, and a short discussion of a future project.

B3.2 Architecture of the BSP400

In this section, the principal components of the BSP400 will be described. Full details of the hardware can be found in Kemna (1990) and in Trienekens and Smouter (1990), and − on the communication protocols − in Melissant and Pelgrom (1990). A global overview of the BSP400's architecture is shown in Figure B3.1. The system consists of a host computer, a network controller, up to 31 units, and a communication bus that connects all units and the network controller.

The *host* computer is necessary to download network parameters into the system,

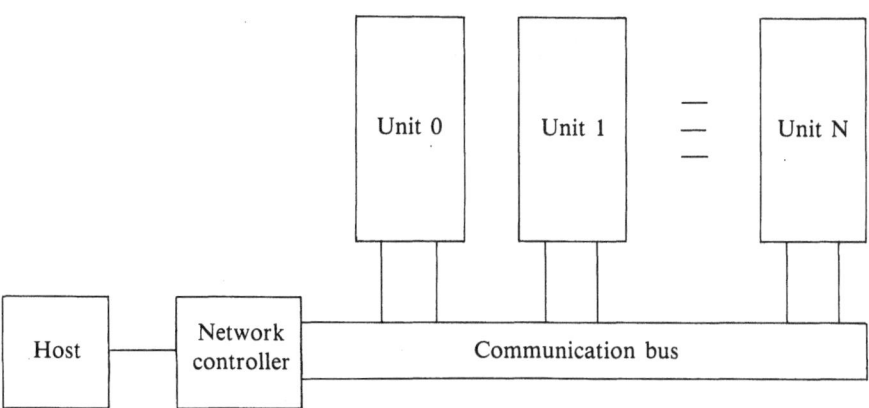

Figure B3.1 Functional overview of BSP400 architecture. Units can be either module boards or I/O boards.

such as network topology, initial weight and activation values. After an initialization phase the host may be disconnected; it serves no crucial role in the ongoing calculations, although it may be used to monitor the system.

The *network controller* has two functions: (i) it synchronizes the calculations by supervising intercommunication and scheduling instruction cycles; (ii) it transfers data and commands between host and units.

Units is the general term for two types of components: (i) *module boards*, each containing 16 processors, and (ii) *I/O boards*, which directly transfer activations to and from the outside world. Direct activation exchange makes the BSP400 suitable for real-world interaction. Module and I/O boards are fully transparent to the network controller. They are completely interchangeable and may be added or removed (if the power is off) with no consequences for the rest of the system.

The module boards
The current implementation of the BSP400 has 25 module boards. Figure B3.2 presents an overview of the functional components on a module board. The board consists of 16 processors plus additional hardware in order to control the dataflow: buses, buffers and control logic.

The module boards are interconnected by the communication bus, which can be decomposed into a DE bus (Data External), a MODE bus, a control bus and a US bus (Unit Select). To control the dataflow of the buses and connected devices on the module, several buffers are used.

An important buffer (or, rather, a FIFO queue) on each module board is the *input activation queue* (IAQ). It mediates the fast external transfer of activation values between units and the relatively slow internal distribution on the module boards. The

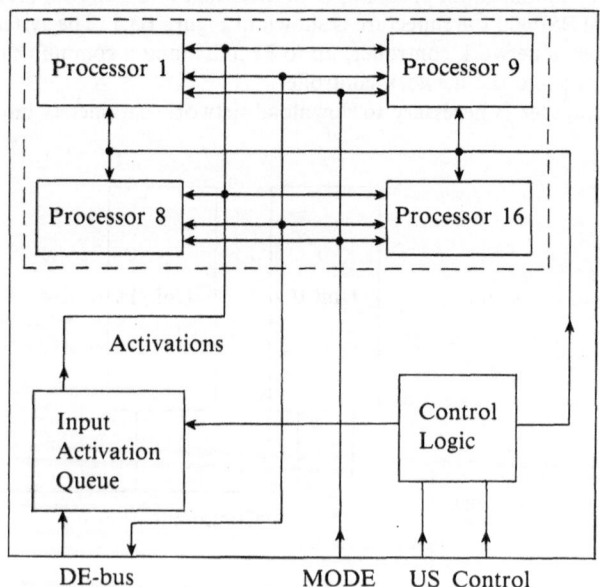

Figure B3.2 Functional view of a module board. The BSP400 consists of 25 such boards. See text for a description of the components.

IAQ's input circuitry is entirely independent of its output circuitry, so after the bus has transferred a set of activation values into the IAQ, it can be used for other purposes. The IAQ is implemented as an asynchronous 64 byte first-in–first-out (FIFO) register. Activation exchange over the bus (DE bus) is several orders of magnitude faster than the internal distribution on a module board of activations and the calculation of activation and weight changes. This implies that a module board never has to 'wait' for external activations.

The module boards are equipped with control logic hardware that is mainly used for synchronization and further transfer of the various operations specified by the network controller (see the next section). Module board selection is achieved by the network controller through the US bus. The nodes select hardware selects the specified processor(s).

The 16 processors form the heart of the parallelism in the BSP400. Each processor can be seen as an LPU (Local Processing Unit), which implements a single node of a neural network. The nodes are implemented as MC68701, 8 bit microprocessors. An MC68701 runs at a 1 MHz clock and contains 2 kbytes of EPROM (UV light erasable, non-volatile program memory) and 128 bytes of RAM. The 2 kbyte EPROM of all nodes may contain the same program although this is not required. In the current implementation the RAM memory is used to store 80 adaptable weights. The remaining 48 RAM bytes are used for parameters and variables like the internal activation of the node. The MC68701 was chosen because it offered programmability, on-chip RAM/ROM, sufficient I/O lines to implement a variety of possible neural chip emulations, and it would allow us to construct and maintain the system at reasonable cost. During the calculation phase the activations are transmitted as single bits. The weights are updated each iteration, so that the system is fully adaptive. Due to the limited RAM space, in this version, only 80 1 byte weights can be stored. This constrains the module boards to receive activations from at most five other units ($5 \times 16 = 80$ input weights).

The network controller

In Figure B3.3, a functional overview of the network controller is shown, indicating the principal dataflow paths. The network controller is implemented by a Z80 processor supplied with 4 kbytes of ROM and 32 kbytes of RAM. The processor runs at 4 MHz.

The network controller is a small self-contained computer with a number of I/O lines and a serial port. The parallel I/O lines are implemented by three parallel I/O chips that drive the communication bus.

The serial communication between the network controller and host is handled by hardware that consists of a serial I/O chip, an RS232 buffer, and a counter timer controller. Commands specified by the host will be translated into MODE words and control signals that will be propagated through the communication bus.

An important part of the network controller is the unit select (US) circuitry which controls a part of the communication bus: the US bus. The US circuitry is a piece of dedicated hardware that enables fast selection of units. The US bus is used to select units either to write or read activations to or from the DE bus. Information about the interunit connections is stored in the connection matrix in the RAM memory of the network controller.

The convergence monitor includes the convergence software. If the convergence software is enabled, the network controller constantly monitors all specified CALM modules (i.e. neural network building blocks) and reports the convergence status to the host in a concise manner. In many cases, the convergence of a CALM module is an

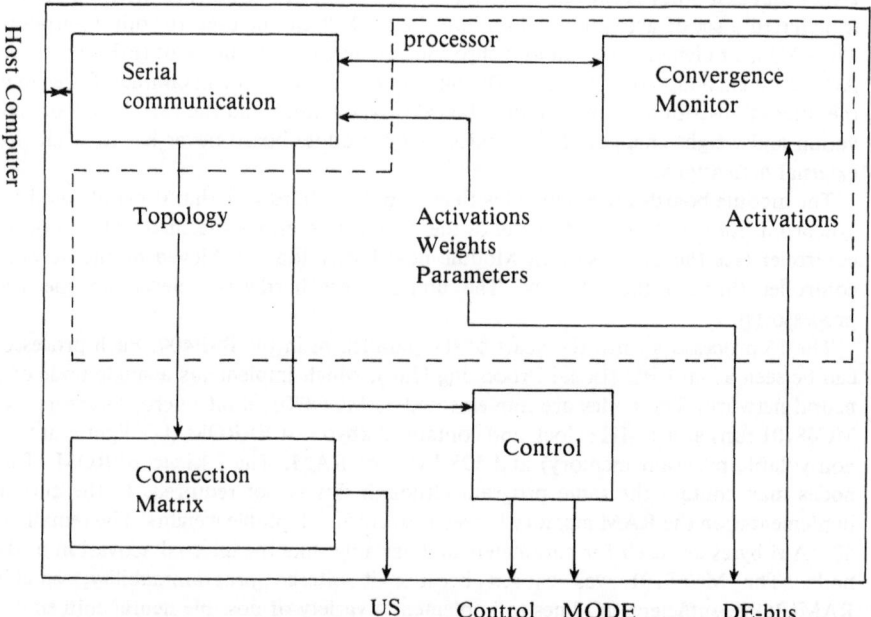

Figure B3.3 Functional view of the network controller. See text for a description of the components.

important condition for certain actions (e.g. motor activation). Convergence checks are executed in parallel (pipelined) with the calculation in the nodes and, therefore, do not slow down the calculation speed.

B3.3 Neural networks on the BSP400

To implement a neural network on the BSP400, all EPROMs of the processors must contain the activation and, where appropriate, learning rules of the network paradigm (e.g. CALM). Then the BSP400 can function as a neurocomputer for that kind of paradigm.

As part of the network controller software a connection matrix is loaded that determines how the units are interconnected. Next, all nodes must be supplied with parameters, such as initial activation state, initial weight values, information specifying which weights are adaptable, learning rate and seed of the pseudo-random number generator. Once a network is initialized in this manner, the calculation process can start.

Like the initialization, the calculation process must be initiated by the host computer. For test purposes a 'single calculation step' command can be used. For real-time processing a 'multiple calculation' command will trigger the BSP400 to continue running on its own. The calculation process can be divided into three phases:

1. *External distribution*. This is the exchange of activations among the units over the connection bus. The units are sequentially prompted to put their activation

values on the DE bus, which in turn are read by (possibly) all units in the system. Activations are read from the DE bus by all selected module boards in parallel and pushed into the input side of their input activation queue (IAQ). The special US circuitry on the network controller and the IAQ ensure fast activation transfer in the external distribution phase.

2. *Internal distribution.* This is the distribution of activations from the IAQ to the individual nodes on a module board. All module boards are operating in parallel. The internal distribution cycle is much slower than the external distribution cycle (see Section B3.5). Because the functioning of the IAQ is fully asynchronous, the network controller can use the communication bus for various purposes during the internal distribution, for example the collection of specified activations to be sent to the I/O boards, or the monitoring of activations for convergence checks. The old activations remain accessible till the end of phase 3.

3. *Calculation of activation and learning rule.* The node software currently implemented contains activation and learning rules of the CALM model. While the nodes are calculating new activations and updating their weights, the network controller may still access old activations of the previous calculation process. At the end of phase 3, the new activations are set to the outputs of the nodes.

B3.4 Discrete implementation of CALM

At present, the memory-resident software in the processors' EPROM only executes operations for the implementation of CALM networks. It is perfectly possible, however, to reprogram the EPROMs to implement some other type of network, subject to the constraint that at most 80 incoming connections per node are allowed coming from four external modules in addition to the connections from nodes on the same module. Also, the activation values must be transferred as single-bit signals, whereas the resolution of the weights is restricted to 1 byte. The latter is due to the limited RAM capacity (128 bytes).

In the original formulation of the CALM algorithm (see Part I), continuous values were used to represent activations and weights. It would seem impossible to implement a model based on categorization through mutually inhibitory connections on the BSP400. In order to compensate for the loss in resolution in the activation values we took recourse to the partial time coding of the activations. This is accomplished by maintaining a 1 byte internal representation of the activation value. If this value exceeds a certain threshold the node (processor) outputs a one, otherwise a zero. Suppose a node receives no input and is currently outputting zeros. If at a certain moment a (constant) input e is administered to the node, it may after a short period start to output ones. The time it takes to switch from outputting zeros to outputting ones depends on the value of e. As soon as it reaches the threshold, the node will start to output ones. As the input activation decreases or is ceased altogether, the internal activation value will drop according to the autonomous decay function. As soon as the internal activation value falls below the threshold, it will start to output zeros again. In this way, the magnitude and duration of the input activation value are partially translated in the number of iterations during which the node outputs ones.

Partial time coding as described above was found to be insufficient for the purpose of implementing categorization based on winner-take-all competition such as in CALM. A problem occurs when two R-nodes receive almost the same amount of activation. As in the continuous version of the CALM module such a deadlock can be broken by adding random noise to the R-nodes. If, however, the input activation is either far below or above the activation threshold the influence of the random noise will not lead to a change in output in the R-nodes. To solve this problem, floating thresholds were introduced for the R-nodes (not for the other nodes). If the E-node is not activated the threshold value of the R-nodes will decrease until it has reached a predefined minimum value, whereas, if it is activated, the threshold will rise until it reaches a predefined maximum value. Suppose that many R-nodes are simultaneously activated; then the E-node will become activated. This will induce rising thresholds in all R-nodes, until only a few are still outputting ones. At this point the E-node activation may turn to zero again, causing the thresholds to drop until more of the R-nodes become activated again. This process can continue for several cycles until one of the R-nodes has won the competition. So, even with binary outputs, the influence of the noise becomes decisive if several R-nodes receive an almost equal activation input. More details on the implementation of floating thresholds can be found in Melissant and Pelgrom (1990).

Because memory space was very limited, the parameters in both the activation rule and the learning rule also had to be rescaled to discrete, hexadecimal values. Some parameter values often used are listed in Table B3.1, which lists the entire contents of the RAM. The loss in precision produced no extra problems in the function of small, discrete CALM modules.

Table B3.1 Contents description of the 128 byte processor RAM.

Parameter name	Description
Weights	Weights of 80 (1 byte) incoming connections
Weight masks	Identification of learning and fixed weights
Activations	Copy of 80 (1 bit) input activations
Old act	Old (1 byte) node activation
New act	New (1 byte) node activation
E-node act	Activation of the E-node (in R-nodes only)
Total act	Total of weighted incoming activations
Noise scale	Scale factor for noisy activations
Scale term	Used in activation rule: $e_i / (\text{Scale term} + e_i)$
Inhibitory scale	Scale factor for inhibitory connections
Fire level	Minimum value of floating activation threshold
Threshold	Floating activation threshold
Threshold step	Step size of floating threshold
ER weight	Weight from E-node to R-nodes
Decay	Autonomous decay in activation rule
Virtual weight	Virtual weight from E-node to lambda
Lambda	Learning parameter
Seed	Seed for the random generator
Variables	Auxiliary variables

B3.5 User interface

Apart from a hardware-reset pushbutton and indicator LEDs on the units and on the network controller, interaction may take place through a program running on the host. After a system reset has been applied to the BSP400, for instance after power-on, a boot program contained in the ROM is started that prepares the network controller to receive network controller software from the host computer. After initialization (see the previous section), various action protocols can be executed by the entire machine. Examples of action protocols are 'Calculate' (either single step or continuously), 'Dump EPROM contents of a node', 'Dump RAM contents of a node', 'Designate output nodes in the system', 'Initialize stimulus', 'Initialize convergence software', 'Reset convergence software', 'Software initiated system reset', and various tests. These actions can be controlled by a menu-driven program, developed in Turbo Pascal.

With a modified version of the current host program it is possible to send batch jobs to the BSP400. This is useful for training purposes. A set of input vectors can be applied together with convergence instructions for the network controller. The network controller reports to the host what is going on in the system and replaces the current input vector by the next vector, for example after convergence.

Direct interaction with the BSP400 through the user interface is at a basic level. The user must manually convert a neural network into a unit connection matrix and fully specify all connections. The BSP400 will, however, be integrated with the neural network environment called MetaNet (described in Appendix B5). MetaNet allows users to create neural networks using a graphical editor. After defining a network (i.e. after constructing a picture of a model on the screen), it can automatically be converted to BSP400 code.

B3.6 Performance analysis

The most important single performance measure for a neurocomputer is the millions of connections per second (MCPS) that are processed. Unfortunately, it is hard to compare different systems on this measure because it is usually not fully clear what is calculated. Other measures for speed are the number of nodes per second (NPS) and the number of iterations per second (IPS), or the iteration time. The BSP400's performance was analysed by investigating the time taken by the various phases of the calculation process described in Section B3.3 (Kemna, 1990). The time to calculate one full iteration on the BSP400 is found by summing the times it takes to complete each of the phases 1 to 3: external distribution time plus internal distribution time plus calculation time (see Table B3.2).

Table B3.2 Decomposition of iteration time into the three principal processing phases.

Phase	Time (ms)	Percentage
External distribution	0.046	0.9
Internal distribution	0.49	9.7
Calculation time	4.5	89.4 +
Total iteration time	5.036	100.0

Table B3.3 Some performance measures for the BSP400. The times do not include learning, but are based only on the execution of the activation rule.

Iterations per second (N_i):	199
Nodes per second (N_n):	79.6×10^3
Connections per second (N_c):	6.4×10^6

The number of iterations per second N_i can be found by dividing one by the iteration time. The maximum number of connections per second N_c is found by multiplying N_i by the total number of connections on the BSP400, which is 400 nodes times 80 connections per node, or 32 000 connections. To find the number of nodes per second, N_n, N_i should be multiplied by 400. The performance figures found with the present design are listed in Table B3.3.

B3.7 Discussion

B3.7.1 Evaluation of the BSP400

From Table B3.2 it becomes clear that in the present implementation the calculation of the activations and weights takes up the greater part, almost 90%, of the iteration time. Any significant improvement in the calculation time will strongly reduce the total iteration time.

The 1 bit transfer of activations may be too limited for certain applications. It is perfectly possible within the current design to multiplex activation bits in order to obtain any desired precision. Each extra bit precision will, however, deteriorate overall performance. More importantly, the very limited 128 byte RAM does not allow for the storage of, say, 80 1 byte activations (in addition to 80 1 byte weights), so that an increase in precision would have to be accompanied by a reduction in the number of incoming connections. RAM space could also be created by giving up some of the precision in the weights. Similar memory problems arise when the BSP400 runs in a virtual mode, where each processor emulates two or more nodes, or where the number of allowed incoming connections is increased.

To give some idea of the performances currently attainable on high-performance implementations of neural networks, Table B3.4 lists some of the fastest implementations known (in January 1991). In terms of 'raw speed', the BSP400 falls far behind the fastest machines. The main reason for the 'low' speed of the BSP400 is the use of the very slow MC68701 processors which run at only 1 MHz. Actually, in biological terms, this speed is not low. The brain is built from very slow components with a 'clock speed' that is perhaps a thousand times lower than that of an MC68701 processor (Feldman, 1981). One of the reasons why the brain works so well is most probably the sheer number of these low-performance elements. In planning the BSP400 project we therefore emphasized the *size* of machine, rather than its speed. A further development of the hierarchical bus structure, which we have called the fractal architecture, allows extensions to a VLSI implementation with a number of processors that is several magnitudes greater than the current design, and with a similar expected increase in speed.

Table B3.4 Six fast neural network implementations (virtual and hybrid implementations). The speed is expressed in millions of connections per second (MCPS). All figures refer to the implementation of NetTalk (Sejnowski and Rosenberg, 1987) using backpropagation in the learning mode.

Machine	Speed (MCPS)	Reference
Cray-2	7	Witbrock and Zagha (1989)
Warp	17	Pomerleau *et al.* (1988)
SX-2	72	Asogawa *et al.* (1988)
Sandy/6	118	Kato *et al.* (1990)
GF11	901	Witbrock and Zagha (1989)
CM-2	1300	Singer (1990)

Nevertheless, faster processors would greatly speed up performance in the current design by decreasing calculation time. Moreover, in this case the internal distribution of activations and calculation could be pipelined with the external distribution. The asynchronous functioning of the IAQ allows the processing of activations at its output side while new activations are still being input at its input side. At the moment we do not make use of this possibility because the MC68701 processors are too slow, so that the external distribution of activations is not a major constraint on performance.

One aspect of the performance cannot easily be improved as long as MC68701 processors are being used, even with scaled-up versions of the BSP400, namely the calculation of the weighted activation input sum. Each of the input activations to a node is multiplied by a stored weight and added to the total input activation. This is a serial process. Also, the calculation of the learning rule and adapting each weight is executed serially. Parallelizing these processes would imply a major improvement in overall performance (see also the example in Section B2.2, Figure B2.1). Nevertheless the limitations of the specific commercial processor have not stood in the way of its usefulness in demonstrating the feasibility of the design and helping define the next generation of the machine.

B3.7.2 Related and future work

We have recently started a new project (Hudson *et al.*, 1990; see also Hudson and Murre, 1991) aimed at extending the BSP400 to 40 000 or more processors. This machine must have an increased flexibility (more paradigms, greater fan-in). It is doubtful whether this extended machine can be completely built from standard, low-cost components. As remarked earlier, the implementation of certain critical parts into special purpose VLSI may greatly increase performance. The experience gained with the BSP400 project will be used to direct the development of this new machine, including VLSI components (Heemskerk *et al.*, 1992). Part of the new project will also be concerned with making fast, dedicated neurocomputers accessible to users through the development of a general, parallel neurosimulator (Murre and Kleynenberg, 1990, 1991; and Appendix B5). It is our conviction that only through a unified approach, combining the integration of computational resources with problem-oriented user interaction, will neurocomputers find applications to a large range of real-world problems. But before presenting this argument in Appendix B5, one more alley to the implementation of neural networks remains to be followed. In Appendix B4, we will take a brief look at the physical implementations and a sketch will be given of a possible implementation of CALM in analog hardware.

Appendix B4

Physical Implementation: Some Notes on CALM in Analog Hardware

B4.1 Motivation for physical implementation

The two previous appendices gave a brief impression of the possibilities of implementing neural networks such as CALM in 'conventional' parallel hardware and hybrid neurocomputers. Though both approaches have yielded promising results, significant improvements may be reached by direct, analog implementations. In a sense, the use of a T800 transputer to implement a number of nodes is overkill; only a tiny fraction of its capabilities is used. The same is true even for the very limited microprocessor used in the BSP400 (MC68701-1). Both are general purpose devices not specifically aimed at serving as nodes in connectionist models. The constituent parts of analog applications, however, such as transistors, resistors, and capacitors, possess many of the necessary characteristics (e.g. different types of non-linearities) to serve as the building blocks of artificial neurons. It seems a waste to allow several unnecessary layers of complex circuitry to carry us up to the level of digital calculation only to model the dynamics of elements that are immediately available at the bottom level.

Moreover, the gain in speed by using direct, analog implementations may range in the order of more than a thousand times, as demonstrated by most of the realized implementations cited below. In addition, the much higher density of the nodes and connections that could be attained in analog VLSI, which even for the case of digital implementations already implies great increases in speed, would allow very large and fast network implementations to become feasible.

Another argument in favour of analog VLSI implementations derives from the lessons to be learned from a confrontation between neural network theory and the physical limits and implementation constraints encountered in such attempts. Among those constraints are obvious ones such as the limited density and bandwidth of interconnections, but also many less apparent aspects, for instance the influence of random variations of components, the effect of signal delays caused by differences in wire length (i.e. the need to synchronize) and of losses of signals due to noise and the volatile implementations of weights. Only analog approaches, either the actual building of analog networks or merely simulating the hardware details, force us to face all of these issues.

Many researchers feel that the constraining factors of hardware implementations of connectionist models are so similar to those found in biological reality that they, therefore, prefer to work on analog implementations of neurophysiological models of, for instance, the early stages of visual processing (Mead and Mahowald, 1988; Taylor, 1990), auditory localization (Lazzaro and Mead, 1989b), pitch perception (Lazzaro and Mead, 1989a), or associative learning in Aplysia (Card and Moore, 1990a,b). The researchers cited here see many specific implementation factors that are common to both hardware and the brain. Some of these implementation factors were also discussed in Appendix B2, when analysing the advantages of modular structures both in the brain and in hardware from an implementation point of view. Modularity is just one of the implementation factors common to biology and analog hardware. Another example is some of the electrical characteristics of EEPROM devices, which have been described (Card and Moore, 1990b) as directly implementing the synaptic weight characteristics of habituation, sensitization, and classical or Pavlovian conditioning. Many implementations forego the digital level altogether, although some of these analog hardware models are actually simulated on digital computers before being implemented in analog VLSI. The latter type of simulation must, however, be distinguished from the virtual implementations in digital hardware in that this simulation is aimed at tracking the behaviour of circuits of analog components implementing neural networks, rather than at directly implementing abstracted neurons in a digital fashion.

B4.2 Some approaches to physical implementation

It is impossible to review all of the existing and proposed techniques for VLSI implementations here. Extensive and up-to-date reviews on this subject can be found in Card and Moore (1989) and in Mead (1989). Instead, we will briefly describe an implementation by Graf *et al.* (1988) to introduce some of the aspects of analog implementations, and to have a point of reference when discussing specific demands made by self-adaptive analog systems. The VLSI implementation reported by Graf *et al.* was developed at AT&T Bell Laboratories using CMOS technology. The integrated circuit consists of an array of 54 simple processing elements. Each element implements a node. The node activations are determined using analog computation. They are fully interconnected with a programmable connection matrix. The connections have ternary values, constant negative, zero, or constant positive. The connections are set by an external host computer.

The aim of Graf *et al.* (1988) was to realize a fast network of reasonably large size, and to explore the practical applications of this approach. A node is implemented by an amplifier unit with two output lines, one for inhibitory and another for excitatory outgoing connections. The chip uses analog computation to calculate the activations, which are represented by voltages. This is done by dividing the voltages by resistors functioning as weights. The resulting currents are summed to represent the input to a processing element. As the authors remark, the summing of currents is accomplished 'for free' on the input lines; merely connecting output wires to an input line results in the desired behaviour. The amplifiers work essentially as threshold elements, so that the activation function of the implemented nodes approaches a step function. The output activations can be considered binary. The programmable connections are implemented by special coupling units. These connect the output of one amplifier to the input of another. Each coupling unit has associated with it 2 bits of RAM. The active (non-zero) bit determines whether the weight on the connection is positive or negative (i.e. which of two output lines from a certain node is being used: the inhibitory or excitatory output line). If both bits are zero the connection is absent (both bits active is apparently not used), and thus the chip uses ternary values for the weights: $\{-L, 0, K\}$. The values of L and K are determined by fixed resistors.

In this implementation, analog computation is used only within the connection matrix: dividing output voltages by resistors and summing the resulting currents. The input and output data to the network are digital, as well as the control signals for the programming of the connections. Input is provided to the network by loading data into 54 1 bit memory cells. Each cell is connected to one amplifier. The loaded values can be used to charge the network with initial voltage levels. During the initialization phase the amplifiers are turned off. For the computation the amplifiers are turned on and the network evolves to a stable state. During this process of network convergence there is no need for any external control or synchronization between the amplifiers. After the circuit has reached a stable state, the output voltage of each amplifier can be read out. The time to reach a stable state typically ranges from 50 up to 600 nanoseconds, depending upon the difficulty of the task.

Although neural networks with binary activations and ternary weights are very limited, a chip as described above can prove useful in a number of simple association and recognition tasks. One of the tasks described by Graf *et al.* (1988) is reminiscent of Steinbuch's (1961) learning matrix, one of the oldest and probably most basic learning networks. As a more ambitious example, the authors also apply their chip to the recognition of digits. More specifically, they use it for the fast line thinning of digits (i.e. reduce all lines to a thickness of one) and for the simple parallel recognition of a number of features. According to reported results the chip functioned well. No spurious states due to malfunctioning were encountered.

Of course, the VLSI implementation by Graf *et al.* does not address the hardest problem: *self-adaptive* or *in situ* learning. Indeed, as Card and Moore (1989) remark, very few implementations of neural networks are self-adaptive. In some cases the connections can be programmed 'off-line', as in the approach above. In most cases, however, the connections are predefined and remain fixed. Card and Moore (1989) argue that the ultimate goal in VLSI implementations is to attain *in situ* learning. But problems will occur when large networks, with possibly millions of connections, must receive a constant supply of weight values. This is necessary in all tasks that involve real-time learning. But even with tasks that only involve periodic (re)learning problems may arise. To enable the fast transfer of weights from external storage to on-chip connections a large number of pins and extensive on-chip wiring will be necessary. This in turn will negatively affect resolution of the circuits and will strongly limit feasible network sizes.

One difficulty that plagues current *in situ* learning implementations is the volatility of the weights. For instance, if capacitors are used to store the weight values of a network they must be provided with a continuous stream of relevant inputs, otherwise the information contained in the connection weights will degrade very quickly. For most real-world applications this is an unsatisfactory situation. Card and Moore (1989) describe several approaches to deal with this problem, most of which are still to be realized.

The difficulties with *in situ* learning apply to a certain extent also to purely optical implementations (e.g. Abu-Mostafa and Psaltis, 1987). Conventional holograms, for instance, are well suited as mass storage sites for weights. They can be easily accessed, but they cannot be retrained. Learning with these devices presents a problem because it demands that the weights be implemented in materials that somehow remain adaptive. A number of successful attempts to develop and apply such materials have recently been reported, for instance the use of photorefractive volume holograms (Psaltis *et al.*, 1990; see also Pepper *et al.*, 1990) or luminescent rebroadcasting devices (e.g. McAulay *et al.*, 1989). The latter devices consist of layers of certain composite coatings. Light of normal frequencies is used to superimpose several images on to a layer. After writing, the material remains 'charged' at the sites that received enough light. By later exposing a layer with infrared light the images can be read out without corrupting the 'charged' sites. In this manner 32 bit bidirectional associative memories (Kosko, 1987) have been implemented. The results were encouraging, although McAulay *et al.* report several drawbacks such as the fact that the output images are of a different frequency than the originals and loss of gain. Several promising alleys are currently being explored (see also Psaltis *et al.*, 1990; Psaltis and Sage, 1988; Treleaven, 1989), and it will be safe to conclude that for the next 5 years or so *in situ* learning will remain one of the important research issues in hardware implementations of neural networks.

Even when VLSI implementations with *in situ* learning become available, problems will remain with the density of connections in models that employ all-to-all connections. As was demonstrated in Appendix B2, virtual and physical implementations of modular networks suffer much less from this problem. The advantage of modular networks applies to most types of implementation techniques, except, perhaps, to optical systems or systems that use non-local activation or learning rules. ART 1 (Grossberg, 1976; Carpenter and Grossberg, 1987, 1988), for example, uses a non-local rule to determine which of the nodes in the F2 layer receives the highest activation input (i.e. a pseudo-neural implementation of winner-take-all competition). It also uses a non-local learning rule. Such non-localities necessitate special purpose circuitry. Indeed, ART 1 may be better suited for implementation in a pipelined, rather than a fully parallel, architecture (e.g. Rao *et al.*, 1990).

B4.3 Outline of a physical implementation of CALM

CALM is well suited for implementation in analog VLSI, because it is modular, uses only fully local activation and learning rules, and has a fully local control structure (see the classification in Section B1.4). Just to illustrate the approach that may be taken to implement CALM in analog hardware, a brief outline of the basic principles of a possible design will be given.

The basic design in analog electronics of a two R-node CALM module is shown in Figure B4.1. The R- and V-nodes, and the A-node, are implemented by an operational amplifier, a number of resistors and a capacitor. This approach is similar to that by Graf *et al.* (1988) and a host of other analog implementations. The output voltage level corresponds to the activation level of the node. When the node gives off inhibitory connections the inverting output of the operational amplifier is used. The short-term retention of an activation, such as used in CALM, can be achieved by a capacitor (see Figure B4.1). Leakage through the capacitor and shunting resistor produces the required activation decay. The intramodular fixed weights in CALM can be implemented by (fixed) resistors. Their values are not critical, since the CALM module has been shown to retain its characteristic functioning under a wide range of weight values. The learning connections can be implemented by one of the techniques discussed in Card and Moore (1989, e.g. p. 157, Figure 8). As discussed above, these types of circuits are still at the focus of research. The voltage-dependent noise generator (VNG) should not present too many problems.

Ready-made analog CALM modules could be put into individual chips. Because the input to a CALM module is limited to the total number of R-nodes in the modules connected to it, the number of input pins could be kept low. The number of output pins would equal the number of R-nodes in the on-chip CALM module. Although this is certainly a workable approach, a great disadvantage is that the machine would become

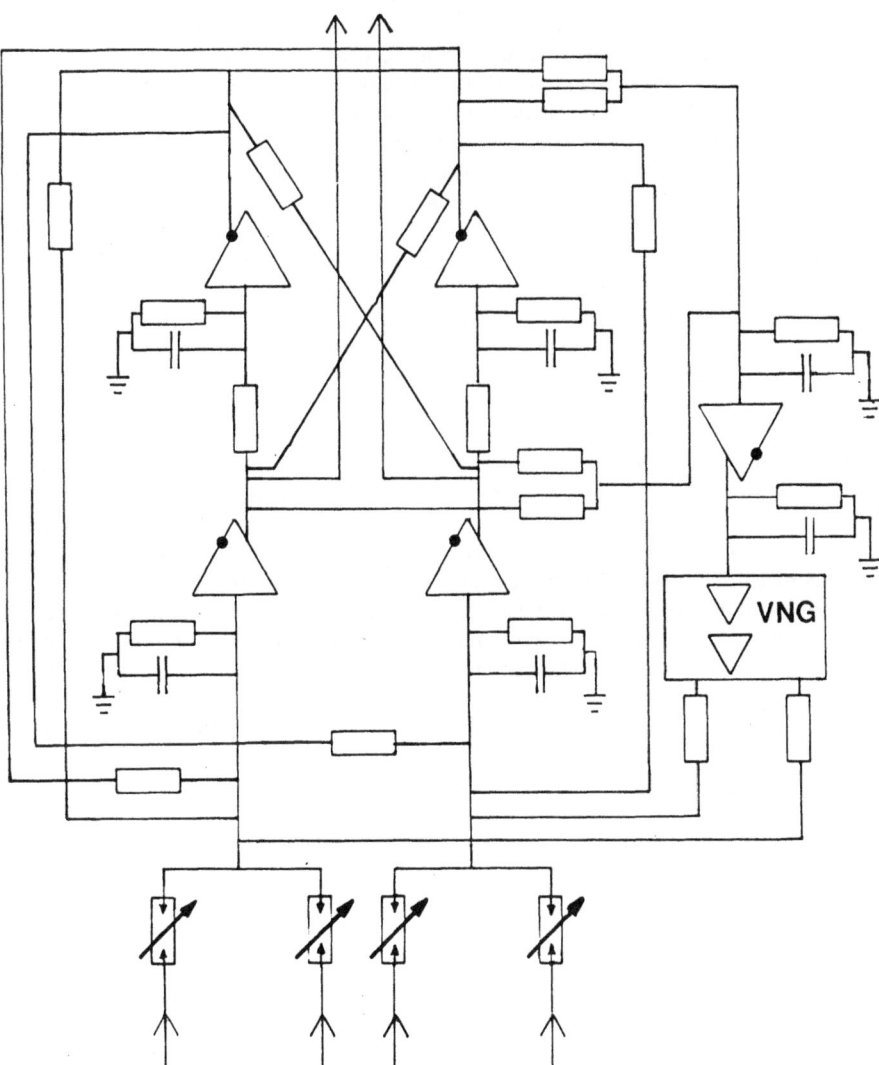

Figure B4.1 Wiring scheme of a two R-node CALM module in analog electronics. The symbols at the very bottom denote variable resistors. VNG indicates a Voltage-dependent Noise Generator.

very inflexible: only a few network topologies could be implemented. Also, one fixed-size CALM module per chip would not be very handy with network topologies that ask for modules in a variety of sizes. Furthermore, provision would have to be made to ensure that a particular chip *A* receives connections from a particular chip *B*. It seems hardly practical to wire each network topology by hand, especially when the networks are very large. To achieve a really scalable design a more practical solution is necessary.

Scalability is an important problem in all parallel computers. Most analog neurocomputers built so far seriously suffer from the inability to extend their design beyond a small number of nodes. In most neurochips marketed, once an implementation has been configured, changing the topology of the neural network model involves the changing of chips and wires by hand. Such a situation should be avoided for most applications. An interesting approach to deal with the scalability issue is the work done by Mueller *et al.* (1989). Their design of an analog neural computer is based on the implementation of network models with a limited connection fan-in to a node (e.g. 64 incoming connections at maximum). The entire design of their machine is somewhat similar to the proposal for an analog implementation of CALM sketched above (or to the BSP400), except that it does not support *in situ* learning. Connection weights only change after the calculation of new values by a host computer followed by serial transfer to each connection. Mueller *et al.* (1989) report a weight transfer time of approximately 2 ms. As remarked earlier, when executed at each iteration such a transfer of the updated weight is very time consuming and strongly deteriorates overall performance.

The most interesting aspect of the design by Mueller *et al.* is the decomposition of the architecture into different types of modules. The machine is built from three types of basic building blocks or modules: 'neuron blocks', 'synapse blocks' and 'switch blocks' for routing. A number of chips of each type is combined into a checkerboard structure which together with a few additional connection lines forms the general architecture of this neurocomputer (see Figure B4.2). A 'neuron chip' contains 16 analog nodes that use a pulse frequency modulation code to transmit output activations. A 'synapse chip' contains 32×16 programmable connections with 6 bit precision (including one sign bit) weight values. Each 'neuron chip' receives connections from two 'synapse chips' which results in a total fan-in of 64 connections per node at maximum. The 'switch chips' contain 32×32 programmable 'crossroads' to connect an arbitrary horizontal line with a vertical line. The switch modules serve to route signals from the output of the 'neuron blocks' to the input of the 'synapse blocks'. By setting the appropriate switches, various network topologies can be realized. The switches themselves could serve as binary connections, almost as in the architecture by Graf *et al.* (1988). Mueller *et al.* (1989) propose, however, to use a two-stage approach that is reminiscent of the BSP400 architecture. The switch blocks (cf. the BSP400 connection matrix) determine the coarse topology of the system, whereas the 5 bit weights in the 'synapse blocks' (cf. the BSP400 weights in RAM) can be used to make the fine distinctions required by most connectionist models. A similar design could be used for an analog implementation of CALM. The division into separate modules for nodes and connections, however, seems to complicate the design unnecessarily.

The general architecture for an analog implementation of CALM would then consist of two principal types of elements: blocks containing nodes and connections (node blocks) and switch blocks. The switch blocks operate in a mode similar to the architecture of Mueller *et al.* In fact, the ensemble of switch blocks forms a sort of reconfigurable analog communication bus with a very high bandwidth. The most optimal placement of the switch blocks is subject to similar concerns as the hierarchical

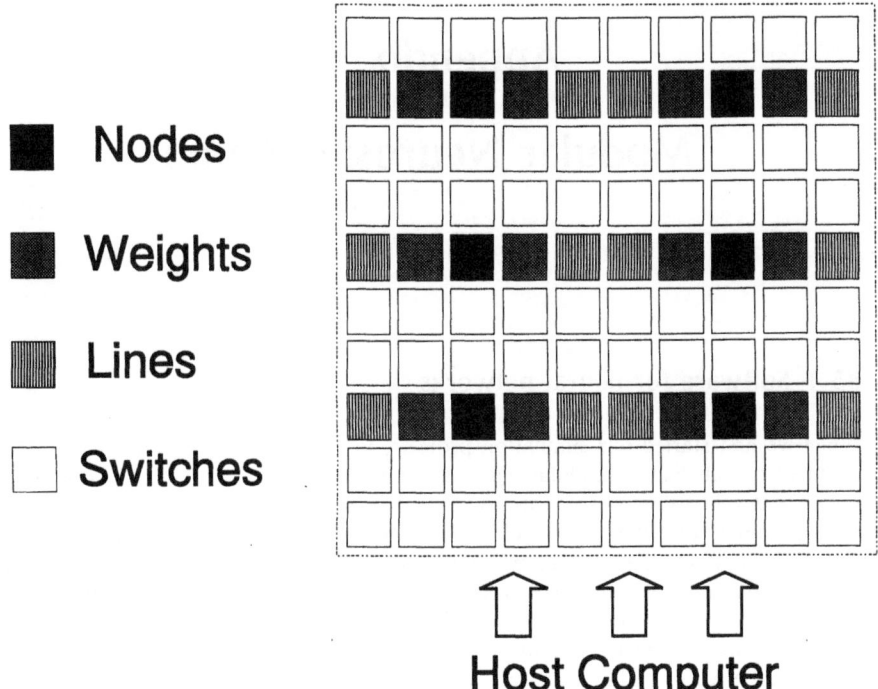

Nodes

Weights

Lines

Switches

Host Computer

Figure B4.2 Layout and general architecture of an analog neurocomputer designed by Mueller *et al.* (1989). Each of the squares is implemented as a separate VLSI chip.

structure of the module interconnect bus in a scaled-up version of the BSP400, especially the fan-in/fan-out density of the nodes and internode pathlength. Multiplexing of connections, however, is impossible in physical implementations, so that the number of wires will be higher. This means that the optimal placement of switch blocks, with regard to routing options for the node-to-node connections to be implemented, may present problems due to lack of physical space. It is simply not possible to support all types of models to be made because, as was demonstrated in Appendix B2, randomly interconnected networks have a comparable degree of connectivity as fully connected models. One way of imposing global limits on connectivity is by restricting the number of modules that can be connected; another is by only allowing locally dispersed connections. A hierarchical connection structure combined with locally dispersed connections may, therefore, be physically possible, while it may be sufficient to support most types of CALM models to be implemented.

The next appendix will focus on ways to interact with neural network models. Three different neural network environments, which were used in the simulations in this work, will be described. It will be argued that for the development of a generally applicable neural network technology, the general neurocomputer, it is crucial that the two problems of integration of computational resources (i.e. neural network hardware) and user interaction (i.e. neural network software) be approached in a unified manner.

Appendix B5

Modular Neurosimulators

B5.1 Software for neural networks

When we first started our simulation projects, in 1988, no available neurosimulator could adequately handle modular types of neural networks such as CALM, so we were forced to develop our own software. As a result of this software implementation effort, in the course of 1988 the *CALM Development System* emerged (Murre and Kleynenberg, 1990). This system, which will be described in Section B5.2, allows users without programming knowledge to construct CALM-based networks. Its primary aim is to support the simulation of psychological experiments. As the need for extra functions grew, the system was gradually extended, which eventually resulted in over 12 000 lines of code with a mild 'spaghetti structure'. This made the routines in the system hard to use in standalone applications, where the internal routines are used directly in another C program. Because of the many graphical possibilities and rich control structure of programming languages (i.e. *for* and *while* loops, and *if...else* constructs), it is often convenient to develop such standalone applications. Therefore, to accommodate the experienced C programmers in our group, a small library of C functions was written, the *CALMLIB*, which will be described briefly in Section B5.3. Although this system has the same basic functionality as the CALM Development System, it has a completely different internal structure (the CALM Development System is array oriented, the CALMLIB is pointer oriented).

As will become clear below, the CALM Development System has a number of limitations, of which we mention the following:

1. It lacks graphical facilities.
2. The control structure of the non-interactive (batch) facility is very limited.
3. It can only run CALM-type networks and it does not allow for experimentation with other networks.
4. The system itself is very large, so that only relatively small networks can be run on a PC.
5. It is not possible to control processing in various kinds of hardware by a single system. The need for this was prompted, among others, by the development of the BSP400 (see Appendix B3).

With the experience gained from the CALM Development System, a new software project has been started. This project has less modest goals; it aims at developing a general, parallel neurosimulator. As a first approximation, the *MetaNet network environment* was developed (Murre and Kleynenberg, 1990, 1991; Murre *et al.*, 1991).

198

To overcome the above-mentioned limitations, it provided the following:

1. A graphical interface, including a graphical editor.
2. A graphical compiler and a network compiler.
3. A higher, textual programming language for the specification of networks and simulations.
4. The option to convert MetaNet code to ANSI C code for compilation to standalone applications.
5. Hardware drivers to buffer the communication in the various processes, possibly on different machines.

The current state of the system, which has already moved away somewhat from the initial system, will be described in Section B5.4. The description is followed by a brief evaluation, concluding that, although MetaNet provides an efficient graphical user interface, it still scores low on extendibility.

In Section B5.5, therefore, an outline for an extended version of MetaNet will be given. This system has a totally rearranged internal structure, which is now based on a concurrent process description. These alterations support the specification of new paradigms. They also complicate the interaction with MetaNet (only at an 'expert' user level), but they have the great advantage that when new models are added, parallelism on different machines is automatically preserved without the need for the explicit synchronization of instructions. Because almost all internal objects are accessible in a single mode of interaction (i.e. direct manipulation), internal consistency is maintained. As in all high-performance applications, simulation environments must compromise between simplicity of use and range of functions. This problem must be addressed by all large software packages. The popular wordprocessor WordPerfect, for instance, has clearly opted for a maximal range of functions. This forms a disadvantage for the limited user, whose progress is often hampered by the confusing multitude of actions. Also, it appears to be very hard for the developers of the package to keep track of internal consistency of operations. In order to deal with this difficult problem in MetaNet, a high, problem-oriented level was introduced. Objects appearing at this level are called POINTs. They can be considered as points of view on the internals of the system. The internal levels are shielded from the user through these interaction tools (POINT stands for Problem-Oriented INteraction Tool), although they remain accessible when necessary.

The appendix closes with a brief section speculating on the possibility of a general neurocomputer, and some of the issues involved in developing such a system. In particular, it is suggested that the internal use of neural networks to achieve a 'neural boot' on large multiprocessors may become important in the near future.

B5.2 The CALM Development System

Designing CALM networks requires thinking in terms of modules. One of the main objectives of the CALM Development System was, therefore, to enable users to create networks on the basis of high-level (modular) specifications. For this purpose an interactive component was developed with a text-based user interface. Some of the features of the interactive component of the CALM Development System are as follows:

● Generating a CALM network on the basis of a high-level specification.

● Saving and loading a network.
● Inspecting a network: activations, connections, modular structure, parameters, node status.
● Changing a network: activations, parameters, node status.
● Initializing a network: activations, connections, parameters.
● Running a network: until convergence, one cycle at a time, i cycles.
● Utilities: setting a new seed for the pseudo-random generator, specifying screen format; help (explanation of commands); and others.

As an example of how a CALM network can be generated in the CALM Development System, the full command dialogue involved in creating a simple two-module network is shown in Listing B5.1. More complex networks, such as used in the simulations in Part II, can be created with only a few more keystrokes.

The interactive component is adequate for designing and testing CALM networks. It does not suffice, however, for extensive simulations of, for instance, complex psychological experiments, where many stimuli are presented in different, predefined orders. In order to support such simulations, the system was extended with a non-interactive,

```
$ calm
 You are now in the CALM Development System; Press h if you need help
calm> m
 Type 'y' if you want to change the module defaults n

    SPECIFY NEW MODEL

How many modules do you want to create? 2

Specify module 1 (press m for menu) m

Options:
  c   CALM module
  i   input module
  o   output module

Specify module 1 (press m for menu) i
What is the number of R-nodes in module 1? 2

Specify module 2 (press m for menu) c
What is the number of R-nodes in module 2? 3

You may now enter the connections between the modules
Always first enter to-module and then enter from-module
Enter 0 for exit

Connection 1
  to-module    -->   2
  from-module  <--   1
Connection 2
  to-module    -->   0

aaa Creating module 1 (Input Module) aaa
aaa Creating module 2 (CALM Module) aaa

Connecting modules
 [2] <- [1]
Connections completed
```

Listing B5.1 Example of the command dialogue involved in creating a simple CALM model in the CALM Development System.

batch-type component. Files with often-used commands can be prepared in advance using a few simple syntactic rules. These command files can then be called and run from the interactive component. The results can either be observed directly, or the results can be written to an output file in a variety of formats for later inspection. After the batch job has ended it can then be evaluated.

In addition to the features listed above, the non-interactive component supports the specification of patterns (stimuli) in terms of node activations and node status, as well as a variety of output formats. Stimuli can be specified with reference to their absolute node address in the model (each module and node has a unique number), or relative to the base node in a module (i.e. the fifth node in module 2). A minimal control structure is available in the form of repetition of command blocks and conditional execution of stimuli.

Despite its obvious limitations, the CALM Development System has been used with success by numerous users over 3 years. It has been implemented on many different computers: VAXs (VMS), PCs (MS-DOS), SUNs (Unix), Amiga computers (Amiga-DOS), and on T800 transputer networks (see Appendix B2).

B5.3 The CALMLIB

The CALMLIB is a small library of approximately 2500 lines of C code. It provides 45 functions and macros supporting an object-oriented approach to network specification and simulation. Only CALM networks can be handled by the present version (Version 2.2) of the library. The library was written to accommodate experienced C programmers, who may prefer developing their own standalone applications rather than using the CALM Development System.

There are a number of good reasons for writing standalone applications. The C syntax and standard libraries offer a wide range of possibilities for experimenting with new network paradigms. Another concern is the output format, which can be adjusted to specific user needs. Graphics routines may be included to show graphs of evolving parameters. An additional library (not discussed here) was developed that contained plotting functions for neural networks, to be used with the CALMLIB. Network parameters may also be dumped in a file for later processing in statistical software packages. User interfaces may be developed especially for the purpose of demonstration or instruction. All of these options were lacking in the CALM Development System. Other reasons involve efficiency considerations. For example, the processing speed may be increased by writing special purpose routines, or the size of code may be diminished by excluding routines that are not used, thus leaving more room for the networks.

There are a number of disadvantages with writing C code. Although an experienced programmer can write a standalone application in a reasonably short time, on certain machines the edit–compile–link–run cycle may take as long as 10 minutes. This, of course, increases development time. The overhead of coding may distract the programmer from the actual task of specifying a simulation. Another disadvantage is that simulations coded directly in C cannot easily be reused by other users, except when the code is very well documented. The batch files in the CALM Development System are all in the same standard format. Once this format is mastered, all simulations can easily be read. These batch files can also be fully transferred between versions of the system running on different types of hardware and operating systems (e.g. from MS-DOS to

VAX-VMS). With applications coded directly in C special care must be taken to ensure portability.

The CALMLIB supports the creation of networks out of CALM modules. Networks may be saved, or loaded from a file. Topology, activations, weights and parameters may be stored selectively. It is not necessary to store an entire network; modules may be stored separately. To facilitate handling of large simulations, files with patterns (pattern batches) may be prepared in advance, and loaded at run time. Selected patterns may then be presented to the network. Calculation of activations and weights, and checking of convergences, can be explicitly specified using a small number of core functions, or the executing of network cycles may be specified by using a single function. Parameters, activations and weights may be reset to their default values by using an

```
#include <stdio.h>                      /* includes necessary header files */
#include <conio.h>
#include "calm.h"

void main ()
{
  int h, i, j, k;                       /* indices                          */
  NETWORKPTR nn;                        /* neural network pointer           */
  BATCHPTR bat;                         /* (pattern) batch pointer          */
  PATTERNPTR ptn;                       /* input module pointer (holds
                                           pattern)                         */

  nn = mdl_load(NULL, "demo.net");      /* load network from file           */
  bat = bat_load("demo.pat");           /* load pattern batch from file     */
  ptn = nmtomdl(nn,"Ptn");              /* get pointer to input module
                                           'Ptn'                            */
  mdl_reset(nn, O_ALL);                 /* initialize act., weights, etc.   */

  for (h = 0; h < 5; h++)               /* do 5 times...                    */
  {
    for (i = 0; i < bat->size; i++)     /* do for entire pattern batch...   */
    {
      mdl_reset(nn,O_INIT);             /* intialize all but the weights    */
      se_ptn(ptn,bat->arr[i]);          /* get pattern i                    */
      for (j = 0; j < 50; j++)          /* for 50 iterations...             */
      {
      calc_act(nn);                     /* calcuate activations             */
      calc_wt(nn);                      /* calculate weights                */
      swap_act(nn);                     /* copy new to old activations      */

      for (k = 0; k < nn->consz; k++)   /* for all modules in nn...         */
        mdl_conv(nn->con[k]->in,j);     /* check and note convergence       */

      gotoxy(1,1);                      /* move cursor to upper-left        */
        mdl_show(nn,O_ALL);             /* show entire model                */

      } /* for j */
    } /* for i */
  } /* for h */

  net_free(nn);                         /* free network                     */
  bat_free(bat);                        /* free pattern batch               */

} /* end main() */
```

Listing B5.2 MS-DOS standalone application with CALMLIB 2.2.

```
T
Demo
2
Ptn
p
2
Mdl1
c
3
Ptn
Mdl1
#
```

Listing B5.3 Network specification file 'demo.net'. T means the beginning of the topology block. The name of the network is Demo. It contains two modules: Ptn is an input (pattern) module, indicated by p, of size 2; Mdl1 is a CALM module, indicated by c, of size 3. Ptn has forward connections to Mdl1. A # marks the end of a block.

```
Demobatch
3
2
1.0  0.0
0.0  1.0
1.0  1.0
```

Listing B5.4 File 'demo.pat' contains a pattern batch with the name Demobatch. The batch consists of three patterns, each of length 2.

initialization function. Functions are provided to show network topology or activations, weights and convergence information in a standard output format.

As an illustration, Listing B5.2 shows how a CALM network and a pattern batch can be loaded and executed. Each pattern is presented for 50 cycles and the presentation of the entire batch is repeated five times. Before a pattern is presented the activations and convergence flags are reset (to zero). The output is shown using the standard output function in the CALMLIB. Note that the size (i.e. number of modules) of the CALM model, as well as the size of the pattern batch (i.e. number of patterns), is not specified in the program. This means that this program can run different models and pattern batches. The only requirements are that the input module is called 'Ptn' (from 'Pattern'), and that the number of elements in the patterns agrees with the size of the input module. Listings B5.3 and B5.4 show an example of a network file and a pattern batch file that could be used in the program of Listing B5.2. The network file only contains a network topology.

B5.4 The MetaNet Network Environment

For reasons set forward repeatedly in this book, we expect that the modular approach to neural networks will gain importance in the near future. We agree with the conclusion of the review by Cohen (1989) that no available neurosimulator offers sufficient support for this approach (see also Murre and Kleynenberg, in prep.). This was one of the reasons that motivated us to develop the MetaNet Network Environment. The only other neurosimulator that we have become aware of, that fully supports the construction of large networks out of smaller modular structures, is the SESAME system (Linden and Tietz, 1992). In contrast with the CALM Development System and the CALMLIB, the MetaNet Network Environment (see also Murre and Kleynenberg, 1990, 1991) was conceived as a *general, parallel neurosimulator*: a system that is consistently extendible with novel network paradigms and parallel hardware. Such a system is faced with a demanding set of problems, because the neural network family includes paradigms with a wide variety of characteristics, and, in addition to this, parallel hardware exhibits a similar range of idiosyncrasies. Mapping a certain neural network onto some type of parallel hardware can, therefore, take many different forms. The situation is aggravated by the fact that users of neurosimulators may range from relative novices to expert users. In particular, the following four problems are major obstacles for the development of a general, parallel neurosimulator:

1. *The representation problem.* How can a problem (i.e. a neural network) be decomposed into a representation that is both manageable and consistent? In our original approach to this problem (Murre and Kleynenberg, 1990), MetaNet was structured using a fixed object hierarchy (i.e. MODEL → MODULE → GROUP, etc.; see the next section). Most recent neurosimulators have adopted a comparable, object-oriented strategy.

2. *The dependency problem.* How can relations between constituent parts of a problem representation be made explicit? The interdependencies between objects in most existing simulators are not handled in a consistent manner. Most systems allow for specific connections to be constructed (e.g. between any two nodes and between layers). Some offer a range of connection schemes that can be applied to only a single kind of object. These connections are of the type typically found in neural networks (i.e. directed connections that have an adaptable, real weight as attribute). In MetaNet, any two objects of the same type can be connected by generalized connections called TRACTs (explained below).

3. *The control problem.* How can a problem representation be converted to a process description without losing potential parallelism? Some readers might wonder whether the control problem applies at all to neural networks. After all, they are neurally inspired structures and should not be subject to some type of central control. In reality, however, many networks are not as 'neural' as desired. For example, the popular backpropagation model (Rumelhart *et al.*, 1986) has a definitely 'non-neural' control structure (see also Appendix B1). In a model with a fully 'neural' control structure, all nodes process their activations in parallel, without any 'sweeps' or other need for explicit synchronization.

We stumbled on the control problem while trying to devise an environment in which users can define networks with novel topologies, parameters and activation and learning dynamics, and that also constrain the order in which the distinguished objects are to be processed (cf. layers in a backpropagation network). A possible solution to this problem is simply to prompt the users to specify such an order. If no special precautions are taken, however, such a linear ordering may result in the partial or total loss of parallelism. Another option is to have the user write out the control structure in some kind of programming language that supports parallelism, for example using some form of concurrent C. For some purposes this approach may be adequate, but because the principal idea of a neurosimulator — or any simulator for that matter – is to free the user from as many distracting details as possible, this can never be a general solution. Specifying an explicitly parallel control structure will be an extra burden for most users that is alien to the actual task: solving a certain problem using neural networks. It should, therefore, be avoided. Our proposed solution to this problem is based on the introduction of several levels of description. A network is converted from a problem-oriented description into a high-level concurrent process description. This code, in turn, is translated into an intermediate code, that is a low-level description of concurrent processes. (See below for a more detailed exposition.)

4. *The machine problem.* How can a process description (i.e. of a certain simulation) be optimally mapped onto a parallel machine? Most parallel neurosimulators deal with machine dependencies in an *ad hoc* manner. This means implementing each specific type of network (e.g. backpropagation) on each available specific type of parallel hardware (e.g. a transputer network), so that the user is still forced to map explicitly a novel type of network onto the supported parallel hardware. Indeed, at present it does not seem feasible to shield the user completely from the machine level, especially if a wide range of parallel machines is to be used. If the concurrency in the process description of a certain model is specified in a sufficiently general format, however, it seems possible to convert (automatically) such intermediate code into a low-level machine-dependent code for many different machines (e.g. into Occam for a transputer network). In Section B5.5, we will propose using a concurrent process description of neural networks as intermediate code. The intermediate code is viewed as an ideal, virtual machine that is emulated by the host hardware. Interpretation of the intermediate code is done by hardware drivers.

B5.4.1 Brief description of the current state of MetaNet

Before discussing a novel approach to problems 3 and 4 by the extended MetaNet Network Environment in Section B5.5, we will first describe the current state of the system. The MetaNet user interface is based on two familiar design concepts: it is menu driven and graphics oriented. Interaction with the system is carried out by use of a mouse. The MetaNet screen is divided into three sections (see Figure B5.1): a menu section to the right, a message section at the bottom, and a large central section for network editing. The edit section is a viewport on an essentially unbounded two-dimensional underlying workspace, in which networks can be constructed. The interface supports scrolling and zooming facilities, allowing the user to view the workspace in any detail, from any viewpoint.

In the current implementation, MetaNet imposes a fixed object hierarchy on a network topology. Four levels of objects are distinguished:

1. NODE
2. GROUP
3. MODULE
4. MODEL

NODEs combine into GROUPs, which assemble into MODULEs. These in turn form MODELs. Constructing a network in MetaNet is done in a top-down fashion, starting at the highest MODEL level and gradually working down to the NODE level. The user first marks a rectangular area of arbitrary size in the workspace. This area becomes a

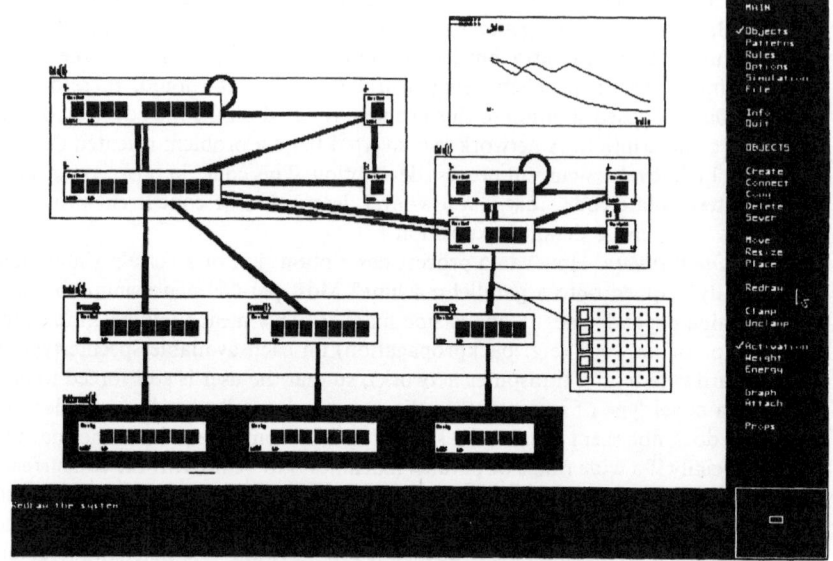

Figure B5.1 The MetaNet screen.

new MODEL. MetaNet allows for the construction and running of multiple models simultaneously, so the user can start by setting up more than one MODEL. Having defined one or more (still empty) MODELs in the workspace, the user can place MODULEs inside the MODELs. If a rectangular area is marked inside a MODEL, the system assumes it to be a new MODULE. The MODULE type forms an important anchoring point in MetaNet. At this level the user can select from standard network architectures (e.g. backpropagation, CALM, Hopfield, etc.) along with learning and control paradigms. If a standard paradigm is chosen, the MODULE will be automatically filled with a standard configuration of GROUPs. In all cases, however, the user may change this configuration (e.g. by deleting GROUPs) or define completely new MODULEs. If a new area is marked inside a MODULE, the system designates it to be a GROUP.

A GROUP consists of NODEs, all of which have to be of the same type. These NODEs are arranged in an $n \times m$ matrix (i.e. a GROUP may be a column, row or full matrix). To assist in network construction a GRID may be superimposed upon the screen, where each of the points indicates a NODE position. In this manner, precise control over the size and the location of NODEs and GROUPs is obtained. Once a number of GROUPs is in place, they can be connected by using the mouse to indicate source and target GROUPs. The set of all NODE-to-NODE CONNECTIONs between two GROUPs is called a TRACT. TRACTs are generalized connections, and actions taken upon TRACTs affect all their constituent CONNECTIONs in a uniform way. By default, the system displays TRACTs, suppressing the multitude of underlying CONNECTIONs. The individual CONNECTIONs can easily be made visible and may be modified, deleted, etc. TRACTs can be formed with different wiring schemes, to be selected from a menu, such as One-to-One, All-to-All, and more complex forms. Because it is possible to connect a NODE in a GROUP to other nodes in the same GROUP (e.g. connecting each node to its adjacent ones), several wiring schemes are incorporated that particularly facilitate the creation of cellular automata, such as the von Neumann and Moore neighbourhoods. In a similar fashion, lateral inhibition can be specified.

The principal objects in MetaNet form two different classes: the unit and the connector class. The unit class consists of MODELs, MODULEs, GROUPs and NODEs. Actions that can be applied to all the members of a class are called its general actions; for the unit class these actions include: create, delete, modify, move, copy, clamp, reset, load, and save. The connector class, consisting of TRACTs and CONNECTIONs, can be manipulated through the following general actions: connect, sever, modify, copy, clamp, and reset. In contrast with general actions, a specific action is only applicable to an individual member of a class. An example of a specific action is changing the activation of a NODE. Together these two classes of functions enable the user to define any two-dimensional network topology, limited only by the amount of memory available to the system.

Due to its modular structure, apart from saving an entire network, parts of it can be saved separately. In this way, libraries of user-defined network components can be constructed and can be loaded into the system at any time.

All objects carrying parameters, like the weight of a CONNECTION or the activation of a NODE, can be inspected and modified. By clicking an object with the mouse, all relevant parameters are displayed in a pop-up window, where they are directly modifiable. MetaNet uses colour scales (or grey scales, depending on the video display unit) to indicate NODE activations or CONNECTION weights. In this way, the dynamics of a running network becomes clearly visible. Other display options are weight matrices, topological maps and state graphs. The system, furthermore, has options such as the possibility of inducing noise on connections or to activate GROUPs of NODEs randomly.

As stated above, the network or parts of it can be saved for later use. MetaNet stores a network's blueprint in a network description file (NDF). The NDF has an ASCII format and may be written to or read by other programs. Also, the NDF can be translated into source code for some standard programming languages. At the moment, only translation into C is supported, but future releases will also include C++ and LISP.

B5.4.2 Evaluation

The UNIT hierarchy in MetaNet functions well, and so does the option to connect UNITs using TRACTs. Starting from scratch, it is possible to create networks rapidly with complicated topological structures. This is particularly true for modular neural networks. The ability to zoom in on any part of the (essentially infinite) workspace and drag around (and delete, 'click open', etc.) anything in view allows for a flexible interaction. The option to suppress detail (e.g. by showing TRACTs rather than the individual connections) further supports the easy handling of networks. Because we have reduced the use of circles, text and dotted lines in the graphical editor, and because we use a very fast graphics kernel, the current implementation of the graphical editor (written in C) gives a fast response, even on modest PCs.

Extending the system with totally novel types of networks, however, presents problems. This is especially true for models that constrain the order in which objects are to be processed, such as layers in backpropagation. Adding such control structure, currently, involves some *ad hoc* programming. Specifying networks using the textual network definition language in the current version of MetaNet (Murre and

Kleynenberg, 1990, 1991) does not blend smoothly with the operation of the graphical interface. The graphical interface, however, does not allow specification of low-level details, such as non-standard convergence checks or activation rules. As will be described below, because of these drawbacks we are no longer developing a textual network definition language. Instead, we now aim at a consistent graphical system, with a high-level *graphical* programming language.

B5.5   Outline of an extension of the MetaNet system

In recent years, many neurosimulators have been developed. A search through the literature revealed no less than 70 neurosimulators (listed in Murre and Kleynenberg, in prep.). Only a small number of these, however, support many different paradigms and many types of parallel hardware. And few neurosimulators (mostly commercial systems) allow the user to extend the system in a consistent manner, for instance by adding a new neural network paradigm or parallel hardware to the available range. In the few consistently extendible neurosimulators, added paradigms often lack the (graphical) interaction possibilities of the 'standard' paradigms, such as revealing the internals of a model (e.g. see the review of three important commercial systems by Cohen (1989)). MetaNet was conceived as a general, parallel neurosimulator. As will be clear from the evaluation, the graphical aspects of the current interface function well, but the system does not yet allow for a consistent extension with novel paradigms and hardware. In other words, we have arrived at an adequate solution for the first two major problems of the four listed at the beginning of Section B5.4, but problems 3 and 4 remain: the control problem and the machine problem. In this section, an extended version of the MetaNet system will be described that approaches these problems by totally revising the internal structure of MetaNet, while leaving most of the external structure unaltered.

In our original approach (Murre and Kleynenberg, 1990), we distinguished between three levels of description: (i) the network definition level, which coincided with both the graphical editor and the textual network definition language; (ii) the intermediate code; and (iii) the machine level. This approach was found to restrict easy extendibility with novel network paradigms, whereas the parallel hardware drivers involved much *ad hoc* programming. The latter occurred primarily with models that had 'non-neural' control structures and, thus, restrict parallelism in an irregular way (see also Section B1.4). In our novel approach we distinguish between the following four levels of description:

1. Problem-oriented description.
2. High-level concurrent process description.
3. Low-level concurrent process description (intermediate code).
4. Machine-dependent description.

Levels 1 to 3 are handled by a graphical editor (see also Figure B5.2), so that all levels of description can be approached in a single style of interaction, closely approximating direct manipulation (e.g. Shneiderman, 1983, 1987). The intermediate code at level 2 allows for a rigorous specification of a problem (i.e. a simulation) in terms of concurrent low-level or atomic processes (see the explanation below). Specifying detailed low-level code is usually done indirectly, through some higher-level language. Here, we have opted for a graphical higher-level language (see below). Because even a higher-level

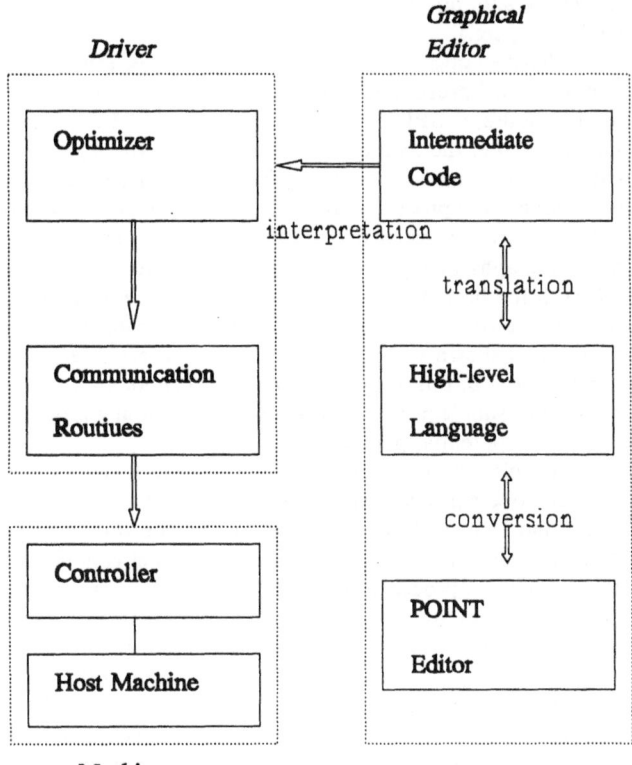

Figure B5.2 Outline of the extended MetaNet system.

concurrent process description contains too many details that are irrelevant for the problem at hand (i.e. getting a neural network to run on some possibly parallel machine), we have included level 1 at which all *objects are problem oriented*. Objects at this level are similar to those found in the current implementation of MetaNet (see Section B5.4).

We have called the level 1 objects POINTs, an abbreviation of Problem-Oriented INteraction Tools. They can be considered as points of view on the lower levels. New POINTs may be defined in terms of level 2 objects, so that the system can be consistently extended with novel paradigms. POINTs are not merely constructed from level 2 objects, but also present a *view* on a collection of level 2 objects. Such a view need not be unique, because several POINTs can be defined that have the same internal, level 2 structure. POINTs do define the way a user *interacts* with such a collection of level 2 objects. Part of a POINT definition includes a list of instructions that tell the POINT how it must be connected to other points (i.e. if the user applies a connection operation). The same collection of level 2 objects may have different connection instructions, thus giving rise to several POINTs. POINTs may differ in appearance (i.e. types of graphs and clickable objects associated with them). In the new approach, POINTs should thus be considered as *points of view* on the higher-level concurrent process description (level 2) and no longer as instances of some fixed object hierarchy (MODEL, MODULE, etc.).

POINTs can be 'clicked open' to reveal level 2 objects. For example, 'clicking open' the output layer of a backpropagation network could reveal such level 2 objects as a GROUP of output nodes, a GROUP of delta nodes, a GROUP of target nodes, and their interconnections (i.e. the data dependencies, see Figure B5.3).

Levels 2 and 3 are based on a concurrent process description. There exists an extensive literature on concurrent processes, and many formal methods have been proposed for representing concurrency problems, for example Petri nets (Petri, 1962; Reisig, 1982), dependency graphs and trace theory (Mazurkiewitz, 1977; Aalbersberg and Rozenberg, 1988). It has been proved that all of these theories can be expressed within the more general framework of process algebra (Mazurkiewitz, 1984). An example of a dependency graph is shown in Figure B5.4. If two processes are connected by an arrow, for example, $A \rightarrow B$, this means that process B cannot start until process A has finished. Based on this simple principle, a representation of any problem can be made that fully captures the potential parallelism.

A computing principle that follows rather closely this approach to representing concurrency is dataflow (see e.g. *Computer* (1982), Gurd (1984) and Gaudiot (1989) for reviews). Denning and Tichy (1990) classify dataflow architectures as the most practical form of MIMD (see Section B1.4) for fine-grained parallel computers known. They expect that the dataflow approach to computation may become practical before the turn of the century. With dataflow, programs are represented as directed graphs in which the nodes (also called actors) are instructions and the arcs are data dependencies. Data are input into one side of the graph and flow through to the output side of the graph (see Figure B5.5). In principle, each node is only executed once. If loops (i.e. iterations) are permitted, however, the pure dataflow principle must be compromised; nodes may be executed more than once. Several ways of 'controlled violation' of the dataflow principle are possible, for instance by only allowing the next iteration to start if all processes in the loop have finished. (This excludes the pipelining of iterations.)

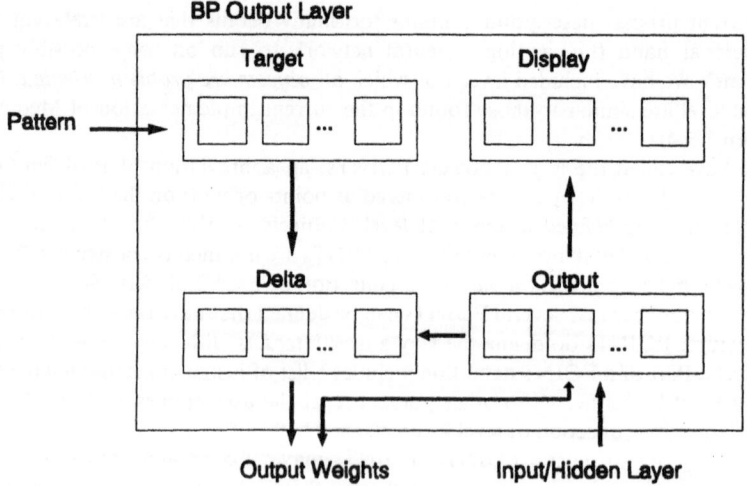

Figure B5.3 A POINT, clicked open to reveal level 2 internal objects.

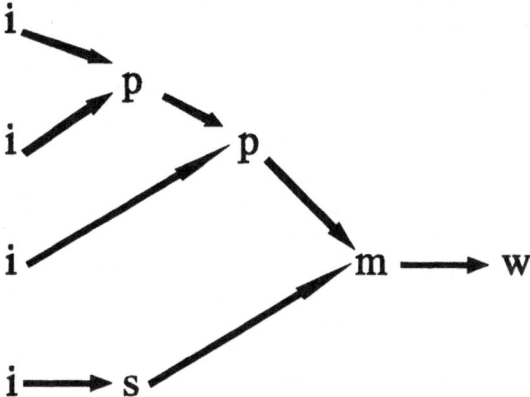

Figure B5.4 Example of a process dependency graph.

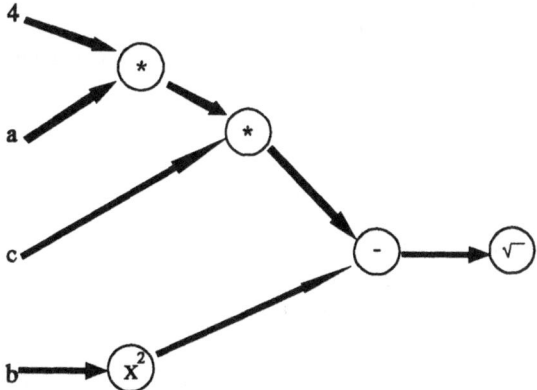

Figure B5.5 An example of a dataflow graph.

Because neural networks are 'non-recursive' structures (i.e. a network does not take another network as input), conversion to a low-level process description in terms of a dependency graph is straightforward. The resulting intermediate code (level 3) can be viewed as an *ideal, virtual machine*. That is, we suppose that some parallel machine exists that precisely fits the process description of the intermediate code. Such an approach to computer design has been called *direct correspondence architecture* (Flynn, 1980; Flynn and Hoevel, 1979; Hoch and Bauer, 1989; Wedig, 1989). The intermediate, level 3 code in MetaNet can be viewed as a *directly executed language* (DEL), the machine code for the virtual, ideal machine. Translation from the level 2 description to the DEL at level 3 is simple. The intermediate code can be characterized as oriented towards the high-level description, rather than to the particularities of some destination machine.

At machine level 4, the ideal, virtual machine must be emulated by the available hardware. This means that the level 3 code must be interpreted for some host machine (e.g. a transputer network). (Note that the expression 'host machine' is used in a different sense than in Appendix B4.) Because the host machine may be a parallel system with a complicated interprocessor communication, the interpretation can take many forms. There may also be an optimization stage in which processor loads and communication overhead are adjusted, so that total iteration time is minimized. For example, on a transputer network with known topology, the interpretation roughly corresponds to an assignment of level 3 processes to transputers. For modular neural networks (but not for fully connected networks, see Appendix B2), it may be advantageous to position the processes in such a manner that communication overhead is minimized (e.g. every node could represent a process). This problem is in a fact a graph K-partitioning problem. For the case where we had only two transputers, the problem would be to divide the neural network into two equal parts such that the number of interconnections between the parts is minimized. This class of problems is well known to be NP-complete, so that some suboptimal algorithm must be used.

Let us summarize the stages through which a network model passes in the extended MetaNet system (see also Figure B5.2):

1. A network model is assembled by the user from a number of POINTs.
2. The model is converted into a high-level description.
3. It is translated into low-level, intermediate code.
4. This code is interpreted and optimized for a certain host machine (by a high-level driver).

Communication with the host is maintained through a set of communication routines (a low-level driver). Process scheduling and synchronization is handled by the machine controller. Further details on the implementation of these stages can be found in Murre and Kleynenberg (in prep.).

Levels 1 to 3 are fully accessible through the graphical editor (e.g. by repeatedly clicking a POINT and its components one finally arrives at the level 3 atomic processes). Through the definition of new POINTs, novel paradigms can be added to the system in a convenient and consistent manner.

B5.6 Towards a general neurocomputer

Although we might well wonder whether a general neurocomputer (GNC) is a possibility, the conclusion, here, is that only a unified approach to user interaction and the integration of computational resources will bring us closer to such a goal (see also Hudson *et al.*, 1990). As was demonstrated in Appendices B2–B4, it is relatively easy to develop hardware for neural networks. It is much harder to apply these systems to a wide range of problems. Only if hardware requirements are gleaned from application results may the development of dedicated neurocomputers be guided in the right direction. A *dedicated* neurocomputer does not imply that such a machine cannot be used in the solution of a wide range of problems. But it remains to be seen whether the range of applications of neurocomputers is large enough to warrant the predicate *general*. The ideas presented in the previous section may be interpreted as aimed towards the development of a GNC. The extended MetaNet system is based on the idea of an ideal, virtual machine. A natural next step would be to develop a machine that approximates this ideal as closely as possible. This is indeed our intention. We have recently started a new project, aimed at developing an extended version of the BSP400, described in

Appendix B3, that must move closer towards the 'ideal neurocomputer'. In a GNC, the role of neural networks need not be limited to being the subject of implementation; there are several ways in which neural networks can be applied *inside* a neurocomputer.

As was described above, an important task for the high-level drivers is to optimize the interpreted level 3 code for balancing processing and communication. Part of this problem can be reduced to a graph K-partitioning problem. If the process dependency graph (of a single iteration) can be partitioned in such a way that the number of connections between partitions is minimized, the communication overhead can be significantly reduced. Recently, a number of authors have applied neural networks to the problems of graph K-partitioning (Peterson and Söderberg, 1989; Van den Bout and Miller, 1990; Van Vliet and Cardon, 1991) and direct load balancing (Schmid, 1990). A logical next step seems to be the implementation of the high-level driver as a graph partitioning neural network, in this way achieving something of a 'neural bootstrap': a neural network is converted to a process dependency graph, which in turn is partitioned using (another) neural network. The partitions can then be assigned to the available processors in an optimal manner (minimizing the total iteration time). An interesting, recent article by Denning and Tichy (1990, p. 1220) hints at a similar application of neural networks in 'ordinary' multicomputers. They remark that neural networks 'are not of direct interest in the numerical computations that predominate in computational science, but they may be of indirect interest for ancillary combinatorial issues such as generating grids and mapping grids to the nodes of a hypercube'. In their opinion, neural networks are best suited for problems with many dissimilar parts. The mapping of a dependency graph onto a processor network is a good example of such a problem. Using neural networks to implement neural networks may become important if multiprocessors or even dedicated neurocomputers become generally available. In that case, users can no longer be expected to map out explicit parallelism in full.

Many other parts of the extended MetaNet system can be implemented by using neural networks or other types of fine-grained parallelism. For example, the user interface itself could be implemented in terms of fine-grained parallel structures (see for instance Van den Bos, 1988). Some of these structures could even be neural networks, used to assist the user in certain tasks (e.g. predicting the next menu item to be chosen). Very little work has been done so far on the application of neural networks to interfaces. Learning neural networks, in particular, would seem to be ideal candidates to support adaptive user interfaces. Another example is the implementation of the conversion, translation and interpretation steps using fine-grained parallelism through parallel graph rewriting.

Bibliography

Aalbergsberg, IJ.J. and G. Rozenberg (1988) 'Theory of traces', *Theoretical Computer Science*, **60**, 1–82.

Abu-Mostafa, Y.S. (1989) 'The Vapnik–Chervonenkis dimension: Information versus complexity in learning', *Neural Computation*, **1**, 312–17.

Abu-Mostafa, Y.S. and D. Psaltis (1987) 'Optical neural computers', *Scientific American*, **256**, (3), 66–73.

Ackley, D.H., G.E. Hinton and T.J. Sejnowski (1985) 'A learning algorithm for Boltzmann machines', *Cognitive Science*, **9**, 147–69.

Ackley, D.H. and M.S. Littman (1990) 'Learning from natural selection in an artificial environment', *Proceedings of the International Joint Conference on Neural Networks, Washington, DC*, vol. 1, pp. 189–193, Hillsdale, NJ: Lawrence Erlbaum.

Ackley, D.H. and M.S. Littman (in prep.) *Evolutionary Reinforcement Learning*.

Aiken, S.W., M.W. Koch and M.W. Roberts (1990) 'A parallel network simulator', *Proceedings of the IEEE ICNN-90, San Diego*, vol. 2, pp. 611–16.

Alexandre, F., F. Guyot and J. Haton (1991) 'The cortical column: A new processing unit for multilayered networks', *Neural Networks*, **4**, 15–25.

Allport, D.A. (1980) 'Patterns and actions', in G.L. Claxton (ed.) *New Directions in Cognitive Psychology*, London: Routledge & Kegan Paul.

Ambros-Ingerson, J., R. Granger and G. Lynch (1990) 'Simulation of paleocortex performs hierarchical clustering', *Science*, **247**, 1344–8.

Anderson, J.A. and E.R. Rosenfeld (eds) (1988) '*Neurocomputing: Foundations of Research*, Cambridge, MA: MIT Press.

Anderson, J.R. and G.H. Bower (1974) *Human Associative Memory*, Washington, DC: Hemisphere.

Asogawa, M., N. Nishi and Y. Seo (1988) 'Network learning on the super computer', *Proceedings of the 36th IPCJ meeting*, pp. 2321–2 (in Japanese, reference cited in Kato *et al.* (1990)).

Atkinson, R.C. and R.M. Shiffrin (1968) 'Human memory: A proposed system and its control processes', in K.W. Spence and T.J. Spence (eds) *The Psychology of Learning and Motivation: Advances in Research and Theory*, vol. 2, pp. 89–195, New York: Academic Press.

Baddeley, A.D. (1986) *Working Memory*, Oxford: Clarendon Press.

Baddeley, A.D. and G.J. Hitch (1974) 'Working memory', in G.H. Bower (ed.) *Recent Advances in the Psychology of Learning and Motivation*, vol. VIII, pp. 47–90, New York: Academic Press.

Bakkers, A.P. (1988) 'Neural controllers and transputers: A survey of the feasibility of neural control systems', *Paper presented at the BIRA Seminar, 18 October 1988, Antwerp, Belgium.*

Ballard, D.H. (1986) 'Cortical connections and parallel processing: Structure and function', *Behavioral and Brain Sciences*, 9, 67–120.

Barlow, H.B. (1981) 'Critical limiting factors in the design of the eye and visual cortex' (The Ferrier Lecture, 1980), *Proceedings of the Royal Society, London*, **B** 212, 1–34.

Barlow, H.B. (1989) 'Unsupervised learning', *Neural Computation*, 1, 295–311.

Barna, G. and K. Kaski (1990) 'Choosing optimal network structure', in B. Widrow and B. Angeniol (eds) *Proceedings of the International Neural Network Conference, INNC-90, Paris*, pp. 890–3, Dordrecht: Kluwer.

Barnsley, M.J. (1988) *Fractals Everywhere*, San Diego, CA: Academic Press.

Barto, A.G., R.S. Sutton and C.W. Anderson (1983) 'Neuronlike adaptive elements that can solve difficult learning control problems', *IEEE Transactions on Systems, Man, and Cybernetics*, **SMC-13**, 834–46.

Baum, E.B. and D. Haussler (1989) 'What size net gives valid generalization?', *Neural Computation*, 1, 151–60.

Bear, M.F. and W. Singer (1986) 'Modulation of visual cortical plasticity by acetylcholine and noradrenaline', *Nature*, 320, 172–6.

Bekhterev, V.M. (1900) 'Demonstration eines Gehirns mit Zerstörung der vorderen und inneren Theile der Hirninde bei Schläferlappens', *Zeitschrift für Neurologie*, **19**.

Belew, R.K. (1989) 'When both individuals and populations search: Adding simple learning to genetic algorithms', in J.D. Schaffer (ed.) *Proceedings of the Third International Conference on Genetic Algorithms and their Applications (ICGA)*, pp. 34–41, San Mateo, CA: Morgan Kaufmann.

Beynon, T. (1988) 'A parallel implementation of the back-propagation algorithm on a network of transputers', *Proceedings of the IEEE ICNN-88, San Diego.*

Blomfield, S. and D. Marr (1970) 'How the cerebellum may be used', *Nature*, 227, 1224–8.

Blumer, A., A. Ehrenfeucht, D. Haussler and M.K. Warmuth (1989) 'Learnability and the Vapnik–Chervonenkis dimension', *Journal of the ACM*, 36, 929–65.

Bounds, D.G. (1987) 'New optimization methods from physics and biology', *Nature*, 329, 215–19.

Box, G.E.P. and G.M. Jenkins (1970) *Time-series Analysis, Forecasting and Control*, San Francisco: Holden Day.

Brady, R.M. (1985) 'Optimization strategies gleaned from biological evolution', *Nature*, 317, 804–6.

Bridgeman, B. (1971) 'Metacontrast and lateral inhibition', *Psychological Review*, 78, 528–39.

Brodsky, M.H. (1990) 'Progress in gallium arsenide semiconductors', *Scientific American*, 262, (2), 56–63.

Brouse, O. and P. Smolensky (1989) 'Virtual memories and massive generalization in connectionist combinatorial learning', *Proceedings of the Eleventh Annual Conference of the Cognitive Science Society*, pp. 380–7, Hillsdale, NJ: Lawrence Erlbaum.

Brown, T.H., P.F. Chapman, E.W. Kairiss and C.L. Keenan (1988) 'Long-term synaptic potentiation', *Science*, 242, 724–8.

Burgess, N. and G. Hitch (1991) 'Towards a network model of the articulatory loop', *Journal of Memory and Language*, in press.

Burnod, Y. (1988) *An Adaptive Neural Network: The cerebral cortex*, Paris: Masson.

Card, H.C. and W.R. Moore (1989) 'VLSI devices and circuits for neural networks', *International Journal of Neural Systems*, **1**, 149–65.

Card, H.C. and W.R. Moore (1990a) 'Silicon models of associative learning in Aplysia', *Neural Networks*, **3**, 333–46.

Card, H.C. and W.R. Moore (1990b) 'Biological learning primitives in analog EEPROM synapses', *Proceedings of IJCNN-90, Washington, DC*, vol. 2, pp. 106–9, Hillsdale, NJ: Lawrence Erlbaum.

Carpenter, G.A. and S. Grossberg (1987) 'Neural dynamics of category learning and recognition: Attention, memory consolidation, and amnesia', in J. Davis, R. Newburgh and E. Wegman (eds) *Brain Structure, Learning, and Memory*, AAAS Symposium Series.

Carpenter, G.A. and S. Grossberg (1988) 'The ART of adaptive pattern recognition by a self-organizing neural network', *Computer*, **21**, 77–88.

Chang, E.I., R.P. Lippmann and D.W. Tong (1990) 'Using genetic algorithms to select and create features for pattern classification', *Proceedings of the IEEE ICNN-90, San Diego*, vol. 3, pp. 747–52.

Changeux, J.P. and A. Danchin (1976) 'Selective stabilisation of developing synapses as a mechanism for the specification of neuronal networks', *Nature*, **264**, 705–12.

Cleeremans, A., D. Servan-Schreiber and J.L. McClelland (1989) 'Finite state automata and simple recurrent networks', *Neural Computation*, **1**, 372–81.

Cofer, C.C. (1967) 'Conditions for the use of verbal associations', *Psychological Bulletin*, **68**, 1–12.

Cohen, H. (1989) 'How useful are current neural network software tools?', *Neural Network Review*, **3**, 102–13.

Computer (1982) 'Special issue on dataflow', *Computer*, **15**, (2).

Conrad, M. (1985) 'On design principles for a molecular computer', *Communications of the ACM*, **28**, 464–80.

Cornsweet, T.N. (1970) *Visual Perception*, New York: Academic Press.

Creutzfeldt, O.D. (1977) 'Generality of the functional structure of the neocortex', *Naturwissenschaften*, **64**, 507–17.

Crick, F. and C. Asanuma (1986) 'Certain aspects of the anatomy and physiology of the cerebral cortex', in J.L. McClelland and D.E. Rumelhart (eds) *Parallel Distributed Processing: Explorations in the microstructure of cognition. Volume 2: Psychological and biological models*, Cambridge, MA: MIT Press.

Damasio, A.R., H. Damasio and G.W. Van Hoesen (1982) 'Prosopagnosia: Anatomic basis and behaviorial mechanisms', *Neurology*, **32**, 331–41.

DARPA (1988) *DARPA Neural Network Study*, Fairfax, VA: AFCEA International Press.

Davis, L. (1987) *Genetic Algorithms and Simulated Annealing*, London: Pitman.

Davis, L. and S. Coombs (1987) 'Optimizing network link sizes with genetic algorithms', in M. Elzas, T. Oren and B.P. Zeigler (eds) *Modelling and Simulation Methodology*, Amsterdam: North-Holland.

Davis, L. and D. Smith (1985) 'Adaptive design for layout synthesis', *Texas Instruments Internal Rep.*, Dallas.

Dawkins, R. (1976) *The Selfish Gene*, Oxford: Oxford University Press.

Dawkins, R. (1986) *The Blind Watchmaker*, New York: Norton.

De Garis, H. (1990a) 'Brain building with GenNets', in B. Widrow and B. Angeniol (eds) *Proceedings of the International Neural Network Conference, INNC-90, Paris*, pp. 1036–9, Dordrecht: Kluwer.

De Garis, H. (1990b) 'Genetic programming: Modular neural evolution for Darwin machines', *Proceedings of the International Joint Conference on Neural Networks, Washington, DC*, vol. 1, pp. 194–7, Hillsdale, NJ: Lawrence Erlbaum.

Denker, J.S., D.B. Schwartz, B.S. Wittner, S.A. Solla, R.E. Howard, L.D. Jackel and J.J. Hopfield (1987) 'Large automatic learning, rule extraction and generalization', *Complex Systems*, 1, 877–922.

Denning, P.J. and W.F. Tichy (1990) 'Highly parallel computation', *Science*, **250**, 1217–22.

Depew, D.J. and B.H. Weber (1989) 'The evolution of the Darwinian research tradition', *Systems Research*, 6, 255–63.

Dodd, N. (1990) 'Optimisation of network structure using genetic algorithms', in B. Widrow and B. Angeniol (eds) *Proceedings of the International Neural Network Conference, INNC-90, Paris*, pp. 693–6, Dordrecht: Kluwer.

Dowsing, R.D. (1988) *Introduction to Concurrency using Occam*, Padstow, Cornwall: T.J. Press.

Ebbinghaus, H. (1885) *Über das Gedächtnis*, Leipzig: Dunker.

Eccles, J.C. (1957) *The Physiology of Synapses*, Berlin: Springer.

Eich, E. (1984) 'Memory for unattended events: Remembering with and without awareness', *Memory and Cognition*, **12**, 105–11.

Elman, J.L. (1988) 'Finding structure in time', *CRL Tech. Rep.* 9901, Center for Research in Language, University of California.

Elman, J.L. (1989) 'Structured representations and connectionist models', *Proceedings of the Eleventh Annual Conference of the Cognitive Science Society*, pp. 17–25, Hillsdale, NJ: Lawrence Erlbaum.

Englander, A.C. (1985) 'Machine learning of visual recognition using genetic algorithms', in J.D. Schaffer (ed.) *Proceedings of the Third International Conference on Genetic Algorithms and their Applications (ICGA)*, pp. 197–201, San Mateo, CA: Morgan Kaufmann.

Ernst, H.P., B. Mokry and Z. Schreter (1990) 'A transputer based general simulator for connectionist models', in R. Eckmiller, G. Hartmann and G. Hauske (eds) *Parallel Processing in Neural Systems and Computers*, pp. 283–6, Amsterdam: North-Holland.

Essen, D. (1985) 'Functional organization of primate visual cortex', in A. Peters and E.G. Jones (eds) *Cerebral Cortex*, vol. 3, pp. 259–329, New York: Plenum Press.

Eysenck, M.W. (1982) *Attention and Arousal*, Berlin: Springer.

Fanty, M.A. (1985) 'Context-free parsing in connectionist networks', *Tech. Rep.* TR174, Computer Science Department, University of Rochester.

Feistel, R. and W. Ebeling (1989) *Evolution of Complex Systems: Self-organization, entropy and development*, Dordrecht: Kluwer.

Feldman, J.A. (1981) 'A connectionist model of visual memory', in G.E. Hinton and J.A. Anderson (eds) *Parallel Models of Associative Memory*, Hillsdale, NJ: Lawrence Erlbaum.

Feldman, J.A. and D.H. Ballard (1982) 'Connectionist models and their properties', *Cognitive Science*, **9**, 1–2.

Ferscha, A. (1990) 'A parallel Boltzmann machine simulator for distributed memory multiprocessor systems', in B. Widrow and B. Angeniol (eds) *Proceedings of the International Neural Network Conference, INNC-90, Paris*, pp. 647–50, Dordrecht: Kluwer.

Flynn, M.J. (1972) *IEEE Transactions on Computing*, **C-21**, 948.

Flynn, M.J. (1980) 'Directions and issues in architecture and language: Language → Architecture → Machine', *Computer*, **13**, (10), 5–22.

Flynn, M.J. and L.W. Hoevel (1979) 'A theory of interpretive architectures: Ideal language machines', *Tech. Rep.* 170, Stanford University.

Fodor, J.A. and Z.W. Pylyshyn (1988) 'Connectionism and cognitive architecture: A critical analysis', *Cognition*, **28**, 3–71.

French, R.M. (1991) 'Using semi-distributed representations to overcome catastrophic forgetting in connectionist networks', *CRCC Tech. Reps* 51-1991, Center for Research on Concepts and Cognition, Indiana University (also submitted to the Cognitive Science Society Conference 1991).

Fu, K.S. (1974) *Syntactic Methods in Pattern Recognition*, New York: Academic Press.

Fukuda, N., Y. Fujimoto and T. Akabane (1990) 'A transputer implementation of toroidal lattice architecture for parallel neurocomputing', *Proceedings of IJCNN-90, Washington, DC*, vol. 2, pp. 43–6, Hillsdale, NJ: Lawrence Erlbaum.

Fukushima, K. (1975) 'Cognitron: A self-organizing multilayered neural network', *Biological Cybernetics*, **20**, 121–36.

Fukushima, K. (1980) 'Neocognitron: A self-organizing neural network model for a mechanism of pattern recognition unaffected by shifts in position', *Biological Cybernetics*, **36**, 193–202.

Fukushima, K. (1988) 'Neocognitron: A hierarchical neural network capable of visual pattern recognition', *Neural Networks*, **1**, 119–30.

Fukushima, K. and S. Miyake (1982) 'Neocognitron: A new algorithm for pattern recognition tolerant of deformations and shifts in position', *Pattern Recognition*, **15**, 455–69.

Fukushima, K., S. Miyake and T. Ito (1983) 'Neocognitron: A neural network model for a mechanism of visual pattern recognition', *IEEE Transactions on Systems, Man, and Cybernetics*, **SMC-13**, 826–34.

Funahashi, K.-I. (1989) 'On the approximate realization of continuous mappings by neural networks', *Neural Networks*, **2**, 183–92.

Gaudiot, J.L. (1989) 'Dataflow machines', in V.M. Milutinović (ed.) *High-level Language Computer Architecture*, Computer Science Press.

Gazzaniga, M.S. (1989) 'Organization of the human brain', *Science*, **245**, 947–52.

Gelernter, D. (1987) 'Programming for advanced computing', *Scientific American*, **256**, (10), 65–71.

Gelernter, D. (1988) 'Getting the job done', *BYTE*, **13**, (12), 301–8.

Gleick, J. (1987) *Chaos: Making a new science*, New York: Viking.

Gluck, M.A. (1991) 'Stimulus generalization and representation in adaptive network models of category learning', *Psychological Science*, **2**, 50–5.

Gluck, M.A. and G.H. Bower (1988) 'Evaluating an adaptive network model of human learning', *Journal of Memory and Language*, **27**, 166–95.

Goldberg, D.E. (1989) *Genetic Algorithms in Search, Optimization, and Machine Learning*, Reading, MA: Addison-Wesley.

Gonzalez, R.C. and M.G. Thomason (1978) *An Introduction to Syntactic Pattern Recognition*, Reading, MA: Addison-Wesley.

Graf, H.P., L.D. Jackel and W.E. Hubbard (1988) 'VLSI implementation of a neural network model', *Computer*, **21**, 41–9.

Graf, P. and G. Mandler (1984) 'Activation makes words more accessible, but not necessarily more retrievable, *Journal of Verbal Learning and Verbal Behavior*, **23**, 553–68.

Graf, P. and D.L. Schacter (1985) 'Implicit and explicit memory for new associations in normal and amnesic subjects', *Journal of Experimental Psychology: Learning, Memory and Cognition*, **11**, 501–18.

Graf, P. and D.L. Schacter (1987) 'Selective effects of interference on implicit and explicit memory for new associations', *Journal of Experimental Psychology: Learning, Memory, and Cognition*, **13**, 45–53.

Gregg, V. (1976) 'Word frequency, recognition and recall', in J. Brown (ed.) *Recall and Recognition*, London: Wiley.

Grossberg, S. (1976) 'Adaptive pattern classification and universal recoding, II: Feedback, expectation, olfaction, and illusions', *Biological Cybernetics*, **23**, 187–202.

Grossberg, S. (1982) *Studies of Mind and Brain: Neural principles of learning, perception, development, cognition, and motor control*, Boston: Reidel.

Grossberg, S. (1984) 'Unitization, automaticity, temporal order, and word recognition', *Cognition and Brain Theory*, **7**, 263–83.

Grossberg, S. (1987a) *The Adaptive Brain. Volume I: Cognition, learning, reinforcement, and rhythm. Volume II: Vision, speech, language, and motor control*, Amsterdam: North-Holland.

Grossberg, S. (1987b) 'Competitive learning: From interactive activation to adaptive resonance', *Cognitive Science*, **11**, 23–63.

Grossberg, S. (1988) 'Nonlinear neural networks: Principles, mechanisms, and architectures', *Neural Networks*, **1**, 17–61.

Grünthal, E. (1939) 'Über das Corpus mamillare und den Korsakowschen Symptomencomplex', *Confinia Neurologica*, **2**.

Gurd, J.R. (1984) 'Fundamentals of dataflow', in F.B. Chambers, D.A. Duce and G.P. Jones (eds) *Distributed Computing*, London: Academic Press.

Guyon, I., I. Poujaud, L. Personnaz, G. Dreyfus, J. Denker and Y. LeCun (1989) 'Comparing different neural network architectures for classifying handwritten digits', *Proceedings of the International Joint Conference on Neural Networks, Washington, DC*, vol. 2, pp. 127–32, New York: IEEE Press.

Halgren, E., N.K. Squires, C.L. Wilson, J.W. Rohrbraugh, T.L. Babb and P.H. Crandall (1980) 'Endogenous potentials generated in the human hippocampal formation and amygdala by infrequent events', *Science*, **210**, 803–5.

Hall, J.F. (1954) 'Learning as a function of word frequency', *American Journal of Psychology*, **67**, 138–40.

Happel, B.L.M. (1990) 'Pattern categorization in multi-module CALM networks: Recognition of handwritten digits', Unpublished Master's thesis, Department of Psychology, Leiden University.

Happel, B.L.M. and J.M.J. Murre (1992) 'Designing modular network architectures using a genetic algorithm', in I. Alexander and J. Taylor (eds) *Proceedings of the International Conference on Artificial Neural Networks, ICANN-92, Brighton*, in press.

Happel, B.L.M., R.H. Phaf, J.M.J. Murre and G. Wolters (1990) 'Categorization in multi-module CALM networks: Recognition of handwritten digits', in B. Widrow and B. Angeniol (eds) *Proceedings of the International Neural Network Conference, INNC-90, Paris*, p. 51 (abstract), Dordrecht: Kluwer.

Harp, S.A., T. Samad and A. Guha (1989) 'Toward the genetic synthesis of neural networks', in J.D. Schaffer (ed.) *Proceedings of the Third International Conference on*

Genetic Algorithms and their Applications (ICGA), pp. 360–9, San Mateo, CA: Morgan Kaufmann.

Hartsuiker, R. (1991) 'Topological self-organization in CALM', *Internal Rep.*, Unit of Experimental and Theoretical Psychology, Leiden University.

Haussler, D. (1990) 'Probably approximately correct learning', *Tech. Rep.* UCSC-CRL-90-16, Computer Research Laboratory, University of California at Santa Cruz.

Hebb, D.O. (1949) *The Organization of Behavior*, New York: Wiley.

Hebb, D.O. (1955) 'Drives and the conceptual nervous system', *Psychological Review*, **62**, 243–54.

Hebb, D.O. (1958) 'Alice in wonderland, or psychology among the biological sciences', in H.F. Harlow and C.N. Woolsey (eds) *Biological and Biochemical Bases of Behavior*, pp. 451–67, Madison, WI: University of Wisconsin.

Heemskerk, J.N.H. and F. Kleer (1989) 'Learning neural networks in digital hardware' (in Dutch: 'Lerende neurale netwerken in digitale hardware'), *Internal Rep.*, Unit of Experimental and Theoretical Psychology, Leiden University.

Heemskerk, J.N.H, J.M.J. Murre, J. Hoekstra, L.H.J.G. Kemna and P.T.W. Hudson (1991) 'The BSP400: a modular neurocomputer, assembled from 400 low-cost microprocessors', in T. Kohonen, K. Mäkisara, O. Simula and J. Kangas (eds) *Artificial Neural Networks: Proceedings of the 1991 International Conference on Artificial Neural Networks (ICANN-91), Espoo, Finland*, pp. 709–14, Amsterdam: Elsevier.

Heemskerk, J.N.H., J.M.J. Murre, A. Melissant, M. Pelgrom and P.T.W. Hudson (1992) 'Mindshape: a neurocomputer concept based on a fractal architecture', in I. Alexander and J. Taylor (eds) *Proceedings of the International Conference on Artificial Neural Networks, ICANN-92, Brighton*, in press.

Hetherington, P.A. (1990) 'Interference and generalization in connectionist networks: Within-domain structure or between-domain correlation?', *Neural Network Review*, **4**, 27–9.

Hetherington, P.A. and M.S. Seidenberg (1989) 'Is there "catastrophic interference" in connectionist networks?', *Proceedings of the Eleventh Annual Conference of the Cognitive Science Society*, pp. 26–33, Hillsdale, NJ: Lawrence Erlbaum.

Hillis, W.D. (1985) *The Connection Machine*, Cambridge, MA: MIT Press.

Hillis, W.D. (1987) 'The Connection Machine', *Scientific American*, **256**, (6), 86–93.

Hinton, G.E. and J.A. Anderson (eds) (1981) *Parallel Models of Associative Memory*, Hillsdale, NJ: Lawrence Erlbaum.

Hinton, G.E. and S.J. Nowlan (1987) 'How learning can guide evolution', *Complex Systems*, **1**, 495–502.

Hoare, C.A.R. (1978) 'Communicating sequential processes', *Communications of the ACM*, **21**, 666–77.

Hoch, J. and C. Bauer (1989) 'The DELtran Project', in V.M. Milutinović (ed.) *High-level Language Computer Architecture*, Computer Science Press.

Hockey, G.R.J., A.W.K. Gaillard and M.G.H. Coles (eds) (1986) *Energetics and Human Information Processing*, Dordrecht: Martinus Nijhoff.

Hoekstra, J. (1990) 'System architecture of a modular neural network using 400 simple processors', *Microprocessing and Micro-programming*, **30**, 257–62.

Hoekstra, J., J.N.H. Heemskerk, A.J. Klaassen, R.H. Phaf, P. Knoppers and P.T.W. Hudson (1990) 'Hardware design concepts for a CALM neural network using 400 simple processors: Architecture and implementation', in B. Widrow and B. Angeniol (eds) *Proceedings of the International Neural Network Conference, INNC-90, Paris*, p. 676 (abstract), Dordrecht: Kluwer.

Hofstadter, D.R. (1985) *Metamagical Themes*, New York: Basic.

Holland, J.H. (1968) 'Hierarchical descriptions of universal spaces and adaptive systems', *Tech. Rep.* ORA Projects 02152 and 08226, Department of Computer and Communication Sciences, University of Michigan, Ann Arbor.

Holland, J.H. (1975) *Adaption in Natural and Artificial Systems*, Ann Arbor, MI: The University of Michigan Press.

Hopcroft, J.E. and J.D. Ullman (1979) *Introduction to Automata Theory, Languages, and Computation*, Reading, MA: Addison-Wesley.

Hopfield, J.J. (1982) 'Neural networks and physical systems with emergent collective computational abilities', *Proceedings of the National Academy of Sciences, USA*, **79**, 2554–8.

Hopfield, J.J. and D.W. Tank (1986) 'Computing with neural circuits: A model', *Science*, **233**, 625–33.

Hornik, K., M. Stinchcombe and H. White (1989) 'Multilayer feedforward networks are universal approximators', *Neural Networks*, **2**, 359–66.

Hubel, D.H. and T.N. Wiesel (1959) 'Receptive fields of single neurons in the cat striate cortex', *Journal of Physiology*, **148**, 574–91.

Hubel, D.H. and T.N. Wiesel (1962) 'Receptive fields, binocular interaction and functional architecture in the cat's visual cortex', *Journal of Physiology*, **160**, 106–54.

Hubel, D.H. and T.N. Wiesel (1963) 'Receptive fields of cells in striate cortex of very young visually inexperienced kittens', *Journal of Neurophysiology*, **26**, 994–1002.

Hubel, D.H. and T.N. Wiesel (1965) 'Binocular interaction in striate cortex of kittens with artificial squint', *Journal of Neurophysiology*, **28**, 1041–59.

Hudson, P.T.W. (1988) 'Construction of a connectionist machine', (in Dutch: 'Bouw van een connectionistische machine'), Project Definition PSYCHON.

Hudson, P.T.W. and J.M.J. Murre (1991) 'A neurosimulator that is consistently extensible with novel network paradigms and parallel hardware', Research Proposal SION 612-322-029, Unit of Experimental and Theoretical Psychology, Leiden University.

Hudson, P.T.W., J.M.J. Murre and R.H. Phaf (1990) 'Neural networks for solving real-world problems in massively parallel hardware: A unified approach to user-interaction and system-integration', Project Definition SION 612-322-018, Unit of Experimental and Theoretical Psychology, Leiden University.

Jacoby, L.L. and M. Dallas (1981) 'On the relationship between autobiographical memory and perceptual learning', *Journal of Experimental Psychology: General*, **110**, 306–40.

James, W. (1890/1950) *Principles of Psychology*, New York: Dover.

Johannet, A., G. Loheac, L. Personnaz, I. Guyon and G. Dreyfus (1987) 'A transputer-based neurocomputer', *Transputer Congress, Grenoble*.

Jordan, M.I (1986) 'Serial order: A parallel distributed processing approach', *Tech. Rep.* 8604, Institute for Cognitive Science, University of California, San Diego.

Jordan, M.I. (1989) 'Serial order: A parallel distributed processing approach', in J.L. Elman and D.E. Rumelhart (eds) *Advances in Connectionist Theory: Speech*, Hillsdale, NJ: Lawrence Erlbaum.

Jordan, M.I. (1990) 'Indeterminate motor skill learning problems', in M. Jeannerod (ed.) *Attention and Performance XIII*, Hillsdale, NJ: Lawrence Erlbaum.

Juliano, S.L., W. Ma and D. Eslin (1991) 'Cholinergic depletion prevents expansion of topographic maps in somatosensory cortex', *Proceedings of the National Academy of Sciences, USA*, **88**, 780–4.

Kaas, J.H., M.M. Merzenich and H.P. Killackey (1983) 'The reorganization of somatosensory cortex following peripheral nerve damage in adult and developing mammals', *Annual Review of Neurosciences*, **6**, 325–56.

Kaas, J.H., R.J. Nelson, M. Sur, C.S. Lin and M.M. Merzenich (1979) 'Multiple representations of the body within the primary somatosensory cortex of primates', *Science*, **204**, 521–3.

Kandel, A. and S.C. Lee (1979) *Fuzzy Switching and Automata: theory and applications*, New York: Crane Russak.

Kandel, E.R. and J.H. Schwartz (1985) *Principles of Neural Science*, New York: Elsevier.

Kanerva, P. (1988) *Sparse Distributed Memory*, Cambridge MA: MIT Press.

Kato, H., H. Yoshizawa, H. Iciki and K. Asakawa (1990) 'A parallel neuro-computer architecture towards billion connection updates per second', *Proceedings of IJCNN-90, Washington, DC*, vol. 2, pp. 47–50, Hillsdale, NJ: Lawrence Erlbaum.

Kauffman, S.A. (1991) 'Antichaos and adaption', *Scientific American*, **265**, (2), 64–70.

Keeler, J.D. (1988) 'Comparison between Kanerva's SDM and Hopfield-type neural networks', *Cognitive Science*, **12**, 299–329.

Kelso, S.R., Ganong, A.H. and T.H. Brown (1986) 'Hebbian synapses in hippo-campus', *Proceedings of the National Academy of Sciences, USA*, **83**, 5326–30.

Kemna, L.H.J.G. (1990) 'The Brain Style Processor. A 400 node hardware implemen-tation of an Artificial Neural Network', *Internal Rep.*, Department of Electrical Engineering, Delft University of Technology.

Kernighan, B.W. and D.M. Ritchie (1988) *The C Programming Language*, 2nd edn, Englewood Cliffs, NJ: Prentice Hall.

Kihlstrom, J.F. (1987) 'The cognitive unconscious', *Science*, **237**, 1445–52.

Kindermann, J. and A. Linden (1990) 'Inversion of neural networks by gradient des-cent', *Parallel Computing*, **14**, 277–86.

Kirkpatrick, S., C.D. Gelatt Jr and M.P. Vecchi (1983) 'Optimization by simulated annealing', *Science*, **220**, 671–80.

Klatt, D.H. (1979) 'Speech perception: A model of acoustic–phonetic analysis and lexical access', *Journal of Phonetics*, **7**, 279–312.

Kleinsmidt, L.J. and S. Kaplan (1963) 'Paired associative learning as function of arousal and interpolated interval', *Journal of Experimental Psychology*, **65**, 190–3.

Kleinsmith, A., M.F. Bear and W. Singer (1987) 'Blockade of "NMDA" receptors disrupts experience-dependent plasticity of kitten striate cortex', *Science*, **238**, 355–8.

Knudsen, E.I., S. Du Lac and S.D. Esterly (1987) 'Computational maps in the brain', *Annual Review of Neurosciences*, **10**, 41–65.

Kohonen, T. (1981) 'Automatic formation of topological maps of patterns in a self-organizing system', *Proceedings of the 2nd Scandinavian Conference on Image Analysis, Espoo, Finland*, pp. 214–20.

Kohonen, T. (1982) 'Self-organized formation of topologically correct feature maps', *Biological Cybernetics*, **43**, 59–69.

Kohonen, T. (1988) 'The "neural" phonetic typewriter', *Computer*, **21**, 11–22.

Kohonen, T. (1989a) *Self-organization and Associative Memory*, 3rd edn, Berlin: Springer.

Kohonen, T. (1989b) 'A self-learning grammar, or "associative memory of the second kind"', *Proceedings of the International Joint Conference on Neural Networks IJCNN-89, Washington, DC*, vol. 1, pp. 1–5, New York: IEEE Press.

Kohonen, T. (1990) 'The self-organizing map', *Proceedings of the IEEE*, **78**, 1464–80.

Koikkalainen, P. and E. Oja (1988) 'Specification and implementation environment for neural networks using communicating sequential processes', *Proceedings of the IEEE ICNN-88, San Diego*, vol. 1, pp. 533–40.

Kosko, B. (1987) 'Competitive adaptive bidirectional associative memories', *Proceedings of ICNN-87*, vol. 2, pp. 759–66.

Kosslyn, S.M., R.A. Flynn, J.B. Amsterdam and G. Wang (1990) 'Components of high-level vision: A cognitive neuroscience analysis and accounts of neurological syndromes', *Cognition*, **34**, 203–77.

Koza, J.R. and M.A. Keane (1990) 'Cart centering and broom balancing by genetically breeding populations of control strategy programs', *Proceedings of the International Joint Conference on Neural Networks, Washington, DC*, vol. 1, pp. 198–201, Hillsdale, NJ: Lawrence Erlbaum.

Krishnaiah, P.R. and L.N. Kanal (eds) (1982) *Handbook of Statistics 2*, Amsterdam: North-Holland.

Kruschke, J.K. (1990) 'ALCOVE: A connectionist model of category learning', Cognitive Science Report 19, Cognitive Science Program, Indiana University.

Kuperstein, M. (1988) 'Neural model of adaptive hand–eye coordination for single postures', *Science*, **239**, 1308–11.

Lazzaro, J. and C.A. Mead (1989a) 'Silicon modeling of pitch perception', *Proceedings of the National Academy of Sciences, USA*, **86**, 9597–601.

Lazzaro, J. and C.A. Mead (1989b) 'A silicon model of auditory localization', *Neural Computation*, **1**, 47–57.

LeCun, Y., B. Boser, J.S. Denker, D. Henderson, R.E. Howard, W. Hubbard and L.D. Jackel (1990a) 'Handwritten digit recognition with back-propagation network', in D. Touretzky (ed.) *Information Processing Systems*, vol. 2, San Mateo, CA: Morgan Kaufmann.

LeCun, Y., L.D. Jackel, H.P. Graf, B. Boser, J.S. Denker, I. Guyon, D. Henderson, R.E. Howard, W. Hubbard and S.A. Solla (1990b) 'Optical character recognition and neural-net chips', in B. Widrow and B. Angeniol (eds) *Proceedings of the International Neural Network Conference, INNC-90, Paris*, pp. 651–5, Dordrecht: Kluwer.

Levelt, W.J.M. (1989) 'De connectionistische mode: Symbolische en subsymbolische modellen van menselijk gedrag', in C. Brown, P. Hagoort and T. Meyering (eds) *Vensters op de geest*, pp. 202–19, Utrecht: Grafiet.

Lewenstein, M. and A. Nowak (1989a) 'Fully connected neural networks with self-control of noise levels', *Physical Review Letters*, **62**, 225–8.

Lewenstein, M. and A. Nowak (1989b) 'Recognition with self-control in neural networks', *Physical Review A* **40**, 4652–64.

Linden, A. and J. Kindermann (1989) 'Inversion of multilayer nets', *Proceedings of the International Joint Conference on Neural Networks, IJCNN-89, Washington, DC*, vol. 2, pp. 425–30, New York: IEEE Press.

Linden, A. and Ch. Tietz (1992) 'SESAME', in I. Alexander and J. Taylor (eds)

Proceedings of the International Conference on Artificial Neural Networks, ICANN-92, Brighton, in press.

Linsker, R. (1986a) 'From basic network principles to neural architecture: Emergence of spatial-opponent cells', *Proceedings of the National Academy of Sciences, USA*, **83**, 7508–12.

Linsker, R. (1986b) 'From basic network principles to neural architecture: Emergence of orientation-selective cells', *Proceedings of the National Academy of Sciences, USA*, **83**, 8390–4.

Linsker, R. (1986c) 'From basic network principles to neural architecture: Emergence of orientation columns', *Proceedings of the National Academy of Sciences, USA*, **83**, 8779–83.

Linsker, R. (1988) 'Self-organization in a perceptual network', *Computer*, **21**, 105–17.

Lippman, R.P. (1989) 'Review of neural networks for speech recognition', *Neural Computation*, **1**, 1–38.

Livingstone, M. and D. Hubel (1988) 'Segregation of form, color, movement, and depth: Anatomy, physiology, and perception', *Science*, **240**, 740–9.

Luria, A.R. (1973) *The Working Brain: An introduction to neuropsychology*, New York: Basic.

McAulay, A.D., J. Wang and C.-T. Ma (1989) 'Optical orthogonal neural network associative memory with luminescent rebroadcasting devices', *Proceedings of IJCNN-89, Washington, DC*, vol. 2, pp. 483–5, New York: IEEE Press.

McClelland, J.L. and D.E. Rumelhart (1981) 'An interactive activation model of context effects in letter perception. Part I: An account of basic findings', *Psychological Review*, **5**, 375–407.

McClelland, J.L and D.E. Rumelhart (eds) (1986a) *Parallel Distributed Processing: Explorations in the microstructure of cognition. Volume 2: Psychological and biological models*, Cambridge, MA: MIT Press.

McClelland, J.L. and D.E. Rumelhart (1986b) 'Amnesia and distributed memory', in J.L. McClelland and D.E. Rumelhart (eds) *Parallel Distributed Processing: Explorations in the microstructure of cognition. Volume 2: Psychological and biological models*, pp. 503–27. Cambridge, MA: MIT Press.

McClelland, J.L. and D.E. Rumelhart (1986c) 'A distributed model of human learning and memory', in J.L. McClelland and D.E. Rumelhart (eds) *Parallel Distributed Processing: Explorations in the microstructure of cognition. Volume 2: Psychological and biological models*, pp. 170–215, Cambridge, MA: MIT Press.

McCloskey, M. and N.J. Cohen (1989) 'Catastrophic interference in connectionist networks: The sequential learning problem', in G.H. Bower (ed.) *The Psychology of Learning and Motivation*, New York: Academic Press.

McGurk, H. and J. MacDonald (1976) 'Hearing lips and seeing voices', *Nature*, **264**, 746–8.

MacLennan, B. (1990) 'Evolution of communication in a population of simple machines', *Tech. Rep.* CS-90-99, Computer Science Department, University of Tennessee.

MacLeod, C. (1989) 'Word context during initial exposure influences degree of priming in word-fragment completion', *Journal of Experimental Psychology: Learning, Memory and Cognition*, **15**, 398–406.

Mandelbrot, B.B. (1982) *The Fractal Geometry of Nature*, San Francisco: Freeman.

Mandler, G. (1980) 'Recognizing: The judgement of previous occurrence', *Psychological Review*, **87**, 252–71.

Maricic, B. and Z. Nikolov (1990) 'GENNET: System for computer aided neural network design using genetic algorithms', *Proceedings of the International Joint Conference on Neural Networks, Washington, DC*, vol. 1, pp. 102–5, Hillsdale, NJ: Lawrence Erlbaum.

Marshall, J.C. (1989) 'An open mind?', *Nature*, **339**, 25–6.

Massaro, D.W. (1986) 'The computer as a metaphor for psychological inquiry: Considerations and recommendations', *Behavior Research Methods, Instruments & Computers*, **18**, 73–92.

Massaro, D.W. (1988) 'Some criticism of connectionist models of human performance', *Journal of Memory and Language*, **27**, 213–34.

Mazurkiewitz, A. (1977) 'Concurrent program schemes and their interpretations', *DAIMI Rep*. PB-78, Aarhus University.

Mazurkiewitz, A. (1984) 'Traces, histories, graphs: Instances of a process monoid', *Lecture Notes in Computer Science*, **176**, 115–33.

Mead, C.A. (1989) *Analog VLSI and Neural Systems*, Reading, MA: Addison-Wesley.

Mead, C.A. and M.A. Mahowald (1988) 'A silicon model of early visual processing', *Neural Networks*, **1**, 91–7.

Melissant, A. and M. Pelgrom (1990) 'Neural networks, a digital implementation', *Internal Rep.*, Unit of Experimental and Theoretical Psychology, Leiden University.

Mewhort, D.J.K. (1990) 'Alice in wonderland, or psychology among the information sciences', *Psychological Research*, **52**, 158–62.

Mewhort, D.J.K. and E.F. Johns (1988) 'Some tests of the interactive-activation model for word identification', *Psychological Research*, **50**, 135–47.

Miller, G.F., P.M. Todd and S.U. Hedge (1989) 'Designing neural networks using genetic algorithms', in J.D. Schaffer (ed.) *Proceedings of the Third International Conference on Genetic Algorithms and their Applications (ICGA)*, pp. 379–84, San Mateo, CA: Morgan Kaufmann.

Milner, B. (1970) 'Memory and the medial temporal regions of the brain', in K.H. Pribram and D.E. Broadbent (eds) *Biology of Memory*, New York: Academic Press.

Mishkin, M. (1978) 'Memory in monkeys severely impaired by combined but by separate removal of amygdala and hippocampus', *Nature*, **273**, 297–8.

Morasso, P., J. Kennedy, E. Antonj, S. Di Marco and M. Dordoni (1990) 'Self-organisation of an allograph lexicon', *International Joint Conference on Neural Networks, Lisbon, March 1990*.

Morse, K.G. (1989) 'In an upscale world', *BYTE*, (August), 222–3.

Mountcastle, V.B. (1978) 'An organizing principle for cerebral function: The unit module and the distributed system', in G.M. Edelman and V.B. Mountcastle (eds) *The Mindful Brain*, Cambridge, MA: MIT Press.

Mueller, P., J. Van der Spiegel, D. Blackman, T. Chiu, T. Clare, J. Dao, C. Donham, T. Hsieh and M. Loinaz (1989) 'A general purpose analog neural computer', *Proceedings of IJCNN-89, Washington, DC*, vol. 2, pp. 177–182, New York: IEEE Press.

Murray, E.A. and M. Mishkin (1984) 'Severe tactual as well as visual memory deficits follow combined removal of the amygdala and hippocampus in monkeys', *Journal of Neuroscience*, **4**, 2565–80.

Murre, J.M.J. (1987) 'Three problems in syntactic pattern recognition', Unpublished Master's Thesis, Utrecht University.

Murre, J.M.J. (1991a) 'Learning in neural networks: A brief overview', (in Dutch: 'Leren in neurale netwerken: Een kort overzicht'), *Informatie*, **33**, 376–88.

Murre, J.M.J. (1991b) 'Transputer implementations of neural networks: An analysis', in T. Kohonen, K. Mäkisara, O. Simula and J. Kangas (eds) *Artificial Neural Networks: Proceedings of the 1991 International Conference on Artificial Neural Networks (ICANN-91), Espoo, Finland*, pp. 1537–40, Amsterdam: Elsevier.

Murre, J.M.J. (1991c) 'Transputers and neural networks: An analysis of implementation constraints and performance', *IEEE Transactions on Neural Networks*, in press.

Murre, J.M.J., J.N.H. Heemskerk, S.E. Kleynenberg and P.T.W. Hudson (1991) 'Integrating the BSP400 neurocomputer with the MetaNet Network Environment', *Proceedings of the International Joint Conference on Artificial Neural Networks, Singapore, 1991*, pp. 1476–80, IEEE No. 91 CH3065-0.

Murre, J.M.J. and S.E. Kleynenberg (1990) 'The MetaNet Network Environment for the development of modular neural networks', in B. Widrow and B. Angeniol (eds) *Proceedings of the International Neural Network Conference, INNC-90, Paris*, pp. 717–20, Dordrecht: Kluwer.

Murre, J.M.J. and S.E. Kleynenberg (1991) 'Extending the MetaNet Network Environment: Process control and machine independence', in T. Kohonen, K. Mäkisara, O. Simula and J. Kangas (eds) *Artificial Neural Networks: Proceedings of the 1991 International Conference on Artificial Neural Networks (ICANN-91), Espoo, Finland*, pp. 545–50, Amsterdam: Elsevier.

Murre, J.M.J. and S.E. Kleynenberg (in prep.) 'Process control and simulation management in general, parallel neurosimulators', *Tech. Rep.*, Unit of Experimental and Theoretical Psychology, Leiden University, in preparation.

Murre, J.M.J., R.H. Phaf and G. Wolters (1989a) 'CALM networks: A modular approach to supervised and unsupervised learning', *Proceedings of the International Joint Conference on Neural Networks, Washington, DC*, vol. 1, pp. 649–56, New York: IEEE Press.

Murre, J.M.J., R.H. Phaf and G. Wolters (1989b) 'Reply to review by T.P. Vogl of "CALM networks: A modular approach to supervised and unsupervised learning"', *Neural Network Review*, **3**, 65–7.

Murre, J.M.J., R.H. Phaf and G. Wolters (1990a) *International Patent Application no.* PCT/NL90/00018.

Murre, J.M.J., R.H. Phaf and G. Wolters (1990b) 'Novelty dependent categorization and learning in CALM modules', in B. Widrow and B. Angeniol (eds) *Proceedings of the International Neural Network Conference, INNC-90, Paris*, pp. 912–15, Dordrecht: Kluwer.

Murre, J.M.J., R.H. Phaf and G. Wolters (1990c) 'CALM: A building block for learning neural networks', in B. Svrecek and J. McRae (eds) *Proceedings of the 1990 Summer Computer Simulation Conference, Calgary*, pp. 781–6, Society for Computer Simulation.

Murre, J.M.J., R.H. Phaf and G. Wolters (1992) 'CALM: Categorizing And Learning Module', *Neural Networks*, **5**, 55–82.

Murrel, G. and J. Morton (1974) 'Word recognition and morphemic structure', *Journal of Experimental Psychology*, **102**, 963–8.

Näätänen, R. (1986) 'The orienting response: A combination of informational and energetical aspects of brain function', in G.R.J. Hockey, A.W.K. Gaillard and M.G.H. Coles (eds) *Energetics and Human Information Processing*, Dordrecht: Martinus Nijhoff.

Nelson, M.E. and J.M. Bower (1990) 'Brain maps and parallel computers', *Trends in Neurosciences*, **13**, 403–8.

Nolfi, S., D. Parisi, G. Vallar and C. Burani (1990) 'Recall of sequences of items by a neural network', in D.S. Touretzky, J.L. Elman, T.J. Sejnowski and G.E. Hinton (eds) *Proceedings of the 1990 Connectionist Summer School*, San Mateo, CA: Morgan Kaufmann.

Nosofsky, R.M. (1985) 'Overall similarity and the identification of separable-dimension stimuli: A choice model analysis', *Perception & Psychophysics*, **38**, 415–32.

Nosofsky, R.M. (1986) 'Attention, similarity, and the identification–categorization relationship', *Journal of Experimental Psychology: General*, **115**, 39–57.

Nosofsky, R.M., S.E. Clark and H.J. Shin (1989) 'Rules and exemplars in categorization, identification, and recognition', *Journal of Experimental Psychology: Learning, Memory, and Cognition*, **15**, 282–304.

Obermayer, K., H. Ritter and K. Schulten (1990a) 'Large-scale simulation of a self-organizing neural network: Formation of a somatotopic map', *International Conference of Parallel Processing in Neural Systems and Computers (ICNC), Düsseldorf, Germany*.

Obermayer, K., H. Ritter and K. Schulten (1990b) 'A principle for the formation of the spatial structure of cortical feature maps', *Proceedings of the National Academy of Sciences, USA*, **87**, 8345–9.

Oja, E. (1982) 'A simplified neuron model as a principal component analyzer', *Journal of Mathematical Biology*, **15**, 267–73.

Ojemann, G.A. (1983) 'Brain organization for language from the perspective of electrical stimulation mapping', *Behavioral and Brain Sciences*, **6**, 189–230.

Partee, B.H., A. Ter Meulen and R.E. Wall (1990) *Mathematical Methods in Linguistics*, Dordrecht: Kluwer.

Penfield, W. and B. Milner (1958) 'Memory deficit produced by bilateral lesions of the hippocampal zone', *Archives of Neurology and Psychiatry*, **74**.

Pepper, D.M., J. Feinberg and N.V. Kukhtarev (1990) 'The photorefractive effect', *Scientific American*, **263**, (4), 34–40.

Perugini, N.K. and W.E. Engeler (1989) 'Neural network learning time: Effects of network and training set size', *Proceedings of the International Joint Conference on Neural Networks, Washington, DC*, vol. 2, pp. 395–401, New York: IEEE Press.

Peterson, C. and B. Söderberg (1989) 'A new method for mapping optimization problems onto neural networks', *International Journal of Neural Systems*, **1**, 3–22.

Petri, C.A. (1962) 'Kommunikation mit Automaten', Schriften des Institutes für Instrumentelle Mathematik, Bonn.

Phaf, R.H. (1986) 'A connectionist model for attention: Restricting parallel processing through modularity', Unpublished Master's Thesis, Leiden University.

Phaf, R.H. (1991) 'Learning in natural and connectionist systems: Experiments and a model', Unpublished Dissertation, Leiden University.

Phaf, R.H. and J.M.J. Murre (1990) 'Cognitie onder de microscoop', in P. Hagoort, C. Brown and T. Meijering (eds) *Vensters op de Geest*, pp. 164–201, Utrecht: Grafiet.

Phaf, R.H., E.O. Postma and G. Wolters (1990a) 'ELAN-1: A connectionist model for implicit and explicit memory tasks', Leiden University Psychological Reports, Unit of Experimental and Theoretical Psychology, EP01-90 (also submitted).

Phaf, R.H., A.H.C. Van der Heijden and P.T.W. Hudson (1990b) 'SLAM: A connectionist model for attention in visual selection tasks', *Cognitive Psychology*, **22**, 273–341.

Phaf, R.H. and G. Wolters (1986) 'Induced arousal and incidental learning during rehearsal', *American Journal of Psychology*, **99**, 341–54.

Phaf, R.H. and G. Wolters (in prep.) 'Spontane aanvulfrequenties voor 168 twee- en drieletterige woordstammen' ('Spontaneous completion frequencies for 168 two- and three-letter word stems'), *Tech. Rep.*, Unit of Experimental and Theoretical Psychology, Leiden University, in preparation.

Phaf, R.H., G. Wolters and M.C.H. Groeneboer (submitted) 'Attention and module-specificity as dissociative factors between implicit and explicit memory'.

Plunkett, K. and C. Sinha (1991) 'Connectionism and developmental theory', *Internal Rep.*, Psychologisk Sriftserie, Aarhus University.

Pomerleau, D.A., G.L. Guscoirla, D.S. Touretsky and H.T. Kung (1988) 'Neural network simulation at Warp speed: How we got 17 million connections per second', *Proceedings of the IEEE ICNN-88, San Diego*, vol. 2, pp. 143–50.

Popper, K.R. (1972) *Objective Knowledge*, Oxford: Oxford University Press.

Prigogine, I. and I. Stengers (1984) *Order Out of Chaos: Man's new dialogue with nature*, New York: Bantam.

Psaltis, D.P., D. Brady and K. Hsu (1990) 'Learning in optical neural computers', *Proceedings of IJCNN-90, Washington, DC*, vol. 2, pp. 72–5, Hillsdale, NJ: Lawrence Erlbaum.

Psaltis, D.P. and J. Sage (1988) 'Advanced implementation technology', in *DARPA Neural Network Study*, pp. 245–86, Fairfax, VA: AFCEA International Press.

Raaijmakers, J.G.W. (1987) 'A formal model for associative memory', in E.E. Roskam and R. Suck (eds) *Progress in Mathematical Psychology*, vol. 1, Amsterdam: North-Holland.

Raaijmakers, J.G.W. and R.M. Shiffrin (1980) 'SAM: A theory of probabilistic search of associative memory', in G. Bower (ed.) *The Psychology of Learning and Motivation*, vol. 14, New York: Academic Press.

Raaijmakers, J.G.W. and R.M. Shiffrin (1981) 'Search of associative memory', *Psychological Review*, **88**, 99–134.

Rabiner, L.R. and B.H. Juang (1986) 'An introduction to Hidden Markov Models', *IEEE ASSP Magazine*, **3**, (1), 4–16.

Rao, A., M.R. Walker, L.T. Clark, L.A. Akers and R.O. Grondin (1990) 'VLSI implementation of neural classifiers', *Neural Computation*, **2**, 35–43.

Ratcliff, R. (1990) 'Connectionist models of recognition memory: Constraints imposed by learning and forgetting functions', *Psychological Review*, **97**, 285–308.

Reale, R.A. and T.J. Imig (1980) 'Tonotopic organization in auditory cortex of the cat', *Journal of Computational Neurology*, **192**, 265–91.

Reisig, W. (1982) *Petrinetze: Eine Einführung*, Berlin: Springer.

Rescorla, R.A. and A.R. Wagner (1972) 'A theory of Pavlovian conditioning: Variations in the effectiveness of reinforcement and non-reinforcement', in A.H. Black and W.F. Prokasy (eds) *Classical Conditioning II: Current research and theory*, pp. 64–99, New York: Appleton-Century-Crofts.

Richardson-Klavehn, A. and R.A. Bjork (1988) 'Measures of memory', *Annual Review of Psychology*, **39**, 475–543.

Ritter, H. and T. Kohonen (1990) 'Learning "semantotopic maps" from context',

Proceedings of the International Joint Conference on Neural Networks, Washington, DC, vol. 1, pp. 23–6, Hillsdale, NJ: Lawrence Erlbaum.

Roberts, B. and W. Li (1988) 'Fukushima's multilayered neural network models', *Neural Network Review*, **2**, 54–60.

Roediger, H.L. (1990) 'Implicit memory: Retention without remembering', *American Psychologist*, **45**, 1043–56.

Roorda-Hrdlicková, V., G. Wolters, B. Bonke and R.H. Phaf (1989) 'Unconscious perception during general anaesthesia demonstrated by an implicit memory task', in B. Bonke, W. Fitch and K. Millar (eds) *Memory and Awareness in Anaesthesia*, pp. 150–6, Amsterdam: Swets & Zeitlinger.

Rosenberg, R.S. (1967) 'Simulation of genetic populations with biochemical properties', Doctoral Dissertation, University of Michigan.

Rueckl, J.G., K.R. Cave and S.M. Kosslyn (1989) 'Why are "What" and "Where" processed by separate cortical visual systems? A computational investigation', *Journal of Cognitive Neuroscience*, **1**, 171–86.

Rumelhart, D.E., G.E. Hinton and R.J. Williams (1986) 'Learning internal representations by error propagation', in D.E. Rumelhart and J.L. McClelland (eds) *Parallel Distributed Processing. Volume 1: Foundations*, Cambridge, MA: MIT Press.

Rumelhart, D.E. and J.L. McClelland (eds) (1986) *Parallel Distributed Processing. Volume 1: Foundations*, Cambridge, MA: MIT Press.

Rumelhart, D.E. and D. Zipser (1985) 'Feature discovery by competitive learning', *Cognitive Science*, **9**, 75–112.

Schacter, D.L. (1987) 'Implicit memory: History and current status', *Journal of Experimental Psychology: Learning, Memory, and Cognition*, **13**, 501–18.

Schacter, D.L. and P. Graf (1986) 'Preserved learning in amnesic patients: Perspectives from research on direct priming', *Journal of Clinical and Experimental Neuropsychology*, **8**, 727–43.

Schacter, D.L. and P. Graf (1989) 'Modality specificity of implicit memory for new associations', *Journal of Experimental Psychology: Learning, Memory and Cognition*, **15**, 3–12.

Schmid, P. (1990) 'The mapping problem: A neural network approach', in B. Widrow and B. Angeniol (eds) *Proceedings of the International Neural Network Conference, INNC-90, Paris*, pp. 274–7, Dordrecht: Kluwer.

Schneider, W. (1987) 'Connectionism: Is it a paradigm shift for psychology?', *Behavior, Research Methods, Instruments, and Computers*, **19**, 73–83.

Schomaker, L.R.B. (1991) 'Simulation and recognition of handwriting movements: A vertical approach to modeling human motor behavior', Unpublished Dissertation, University of Nijmegen.

Scoville, W.B. (1954) 'The limbic system in man', *Journal of Neurosurgery*, **11**.

Scoville, W.B. and B. Milner (1957) 'Loss of recent memory after bilateral hippocampal lesions', *Journal of Neurology, Neurosurgery, and Psychiatry*, **120**, 11–21.

Sejnowski, T.J. and C.R. Rosenberg (1986) 'NETtalk: A parallel network that learns to read aloud', Johns Hopkins University Electrical Engineering and Computer Science Report TR JHU/EECS-86/01.

Sejnowski, T.J. and C.R. Rosenberg (1987) 'Parallel networks that learn to pronounce English text', *Complex Systems*, **1**, 145–68.

Shallice, T. (1988) 'Specialisation within the semantic system', *Cognitive Neuropsychology*, **5**, 133–142.

Shallice, T., P. McLeod and K. Lewis (1985) 'Isolating cognitive modules with the dual-task paradigm: Are speech perception and production separate processes?', *The Quarterly Journal of Experimental Psychology*, **37A**, 507–32.

Shepard, R.N. (1957) 'Stimulus and response generalization: A stochastic model, relating generalization to distance in psychological space', *Psychometrika*, **22**, 325–45.

Shepard, R.N. (1987) 'Toward a universal law of generalization for psychological science', *Science*, **237**, 1317–23.

Shimamura, A.P. (1986) 'Priming effects in amnesia: Evidence for a dissociable memory function', *Quarterly Journal of Experimental Psychology*, **38A**, 619–44.

Shimamura, A.P. and L.R. Squire (1989) 'Impaired priming of new associations in amnesia', *Journal of Experimental Psychology: Learning, Memory, and Cognition*, **15**, 721–8.

Shneiderman, B. (1983) 'Direct manipulation: A step beyond programming languages', *Computer*, **16**, 57–69.

Shneiderman, B. (1987) *Designing the User Interface: Strategies for effective human–computer interaction*, Reading, MA: Addison-Wesley.

Shors, T.J., T.B. Seib, S. Levine and R.F. Thompson (1989) 'Inescapable versus escapable shock modulates long-term potentiation in the rat hippocampus', *Science*, **244**, 224–6.

Simon, J.C., E. Backer and J. Sallentin (1982) 'A unifying viewpoint on pattern recognition', in P.R. Krishnaiah and L.N. Kanal (eds) *Handbook of Statistics 2*, Amsterdam: North-Holland.

Singer, A. (1990) 'Exploiting the inherent parallellism of artificial neural networks to achieve 1300 million interconnects per second', in B. Widrow and B. Angeniol (eds) *Proceedings of the International Neural Network Conference, INNC-90, Paris*, pp. 656–60, Dordrecht: Kluwer.

Smith, A.W. and D. Zipser (1989) 'Learning sequential structure with the real-time recurrent learning algorithm', *International Journal of Neural Systems*, **1**, 125–31.

Sokolov, E.N., (1960) 'Neuronal models and the orienting reflex', in M.A. Brazier (ed.) *The Central Nervous System and Behaviour*, New York: Josiah Macy Jr Foundation.

Sokolov, E.N., (1966) 'Orienting reflex as information regulator', in A.N. Leontiev, A.R. Luria and S. Smirnov (eds) *Psychological Research in the USSR*, Moscow: Progress.

Sokolov, E.N., (1975) 'The neuronal mechanisms of the orienting reflex', in E.N. Sokolov and O.S. Vinogradova (eds) *Neuronal Mechanisms of the Orienting Reflex*, Hillsdale, NJ: Lawrence Erlbaum.

Solla, S.A. (1989) 'Learning and generalization in layered neural networks: The contiguity problem', in L. Personnas and G. Dreyfus (eds) *Neural Networks: From models to applications*, pp. 168–77, Paris: IDSET.

Spears, W.M. and K.A. De Jong (1990) 'Using neural networks and genetic algorithms as heuristics for NP-complete problems', *Proceedings of the International Joint Conference on Neural Networks, Washington, DC*, vol. 1, pp. 118–121, Hillsdale, NJ: Lawrence Erlbaum.

Spitzer, H., R. Desimone and J. Moran (1988) 'Increased attention enhances both behavioral and neuronal performance', *Science*, **240**, 238–40.

Squire, L.R. (1986) 'Mechanisms of memory', *Science*, **232**, 1612–19.

Squire, L.R. and N.J. Cohen (1984) 'Human memory and amnesia', in J. McGaugh, G. Lynch and N. Weinberger (eds) *Proceedings of the Conference on the Neurobiology of Learning and Memory*, pp. 3–64, New York: Guilford Press.

Squire, L.R. and S. Zola-Morgan (1988) 'Memory brain systems and behavior', *Trends in Neurosciences*, **4**, 170–5.

Stanfill, D. and D. Waltz (1986) 'Toward memory-based reasoning', *Communications of the ACM*, **29**, 1213–28.

Stein, R.M. (1988) 'T800 and counting', *BYTE*, **13**, (12), 287–96.

Steinbuch, K. (1961) 'Die Lernmatrix', *Kybernetik*, **1**, 36–45.

Stinchcombe, M. and H. White (1989) 'Universal approximation using feedforward networks with non-sigmoid hidden layer activation function', *Proceedings of the International Joint Conference on Neural Networks, Washington, DC*, vol. 1, pp. 613–17, New York: IEEE Press.

Sugimoto, D., Y. Chikada, J. Makino, T. Ito, T. Ebisuzaki and M. Umemura (1990) 'A special-purpose computer for gravitational many-body problems', *Nature*, **345**, 33–5.

Swaine, M. (1989) 'Two early neural net implementations', *Dr Dobb's Journal of Software Tools*, **16**, (11), 124–31.

Szentágothai, J., (1975) 'The "module-concept" in cerebral cortex architecture', *Brain Research*, **95**, 475–96.

Taylor, J.G. (1990) 'A silicon model of vertebrate retinal processing', *Neural Networks*, **3**, 171–8.

Todd, P.M. (1989a) 'Review of "A self-learning musical grammar, or 'associative memory of the second kind'" by T. Kohonen', *Proceedings of the International Joint Conference on Neural Networks, IJCNN-89 Washington, DC, Neural Network Review*, vol. 3, pp. 114–15 (with a response from the author).

Todd, P.M. (1989b) 'A connectionist approach to algorithmic composition', *Computer Music Journal*, **13**, 27–43.

Todd, P.M. and D.G. Loy (eds) (1991) *Music and Connectionism*, Cambridge, MA: MIT Press.

Treleaven, P.C. (1989) 'Neurocomputers', *International Journal of Neurocomputing*, **1**, 4–31.

Trienekens, W.P. and F. Smouter (1989) '400 nodes neural network. A design of a network with 400 processors for neural network simulation', (in Dutch: '400 nodes neuraal netwerk. Een ontwerp van een netwerk met 400 processoren ten behoeve van neurale netwerksimulaties'), Unpublished Master's Thesis, Delft University of Technology.

Tulving, E. (1972) 'Episodic and semantic memory', in E. Tulving and W. Donaldson (eds) *Organisation of Memory*, pp. 381–403, New York: Academic Press.

Tulving, E. (1983) *Elements of Episodic Memory*, Oxford: Oxford University Press.

Tulving, E. (1985) 'How many memory systems are there?', *American Psychologist*, **40**, 385–98.

Tulving, E. and D.L. Schacter (1990) 'Priming and human memory systems', *Science*, **247**, 301–6.

Tunturi, A.R. (1950) 'Physiological determination of the arrangement of the afferent connections to the middle ectosylvian auditory area in the dog', *American Journal of Physiology*, **162**, 489–502.

Tunturi, A.R. (1952) 'The auditory cortex of the dog', *American Journal of Physiology*, **168**, 712–17.

Uit den Boogaart, P.C. (ed.) (1975) *Woordfrequenties in geschreven en gesproken Nederlands* (in English: *Word Frequencies in Spoken and Written Dutch*), Utrecht: Oosthoek, Scheltens en Holkema.

Vailliant, L.G. (1984) 'A theory of the learnable', *Communications of the ACM*, **27**, 1134–42.

Valdés, R. (1991) 'What is BioComputing?', *DrDobb's Journal*, **16**, (4), 46 and 108–9.

Van den Bos, J. (1988) 'Abstract Interaction Tools: A language for user interface management systems', *ACM Transactions on Programming Languages and Systems*, **10**, 215–47.

Van den Bout, D.E. and T.K. Miller III (1989) 'Graph partitioning using annealed neural networks', *Proceedings of IJCNN-89, Washington, DC*, vol. 1, pp. 521–8, New York: IEEE Press.

Van den Bout, D.E. and T.K. Miller III (1990) 'Graph partitioning using annealed neural networks', *IEEE Transactions on Neural Networks*, **1**, 192–203.

Van den Bout, T. (1990) 'Transputers & neural networks', (In Dutch: 'Transputers & neurale netwerken'), Unpublished Master's Thesis, Leiden University.

Van Vliet, R. and H. Cardon (1991) 'Combining a graph partitioning and a TSP neural network to solve the MTSP', in T. Kohonen, K. Mäkisara, O. Simula and J. Kangas (eds) *Artificial Neural Networks: Proceedings of the 1991 International Conference on Artificial Neural Networks (ICANN-91), Espoo, Finland*, pp. 157–162, Amsterdam: Elsevier.

Vapnik, V.N. and A. Chervonenkis (1971) 'On the uniform convergence of relative frequencies of events to their probabilities', *Theory of Probability and its Applications*, **16**, 264–80.

Vogl, T.P., (1989) 'Review of CALM networks: A modular approach to supervised and unsupervised learning', *Neural Network Review*, **3**, 64–5.

Von der Malsburg, C. (1973) 'Self-organization of orientation sensitive cells in the striate cortex', *Kybernetik*, **14**, 85–100.

Von Neumann, J. (1958) *The Computer and the Brain*, New Haven, CT: Yale University Press.

Waibel, A. (1989) 'Modular construction of time-delay neural networks for speech recognition', *Neural Computation*, **1**, 39–46.

Walley, R.E. and T.D. Weiden (1973) 'Lateral inhibition and cognitive masking: A neuropsychological theory of attention', *Psychological Review*, **80**, 284–302.

Warren, R.M. (1984) 'Perceptual restoration of obliterated sounds', *Psychological Bulletin*, **96**, 371–83.

Warrington, E.K. and L. Weiskrantz (1968) 'A new method of testing long-term retention in amnesic patients', *Nature*, **217**, 972–4.

Warrington, E.K. and L. Weiskrantz (1970) 'Amnesic syndrome: consolidation or retrieval?', *Nature*, **228**, 628–30.

Weber, B.H., D.J. Depew, C. Dyke, S.N. Salthe, E.D. Schneider, R.E. Ulanovicz and J.S. Wicken (1986) 'Evolution in thermodynamic perspective: an ecological approach', *Biology and Philosophy*, **4**, 373–405.

Wedig, R.G. (1989) 'Direct correspondence architectures: Principles, architectures, and design', in V.M. Milutinović (ed.) *High-level Language Computer Architecture*, Computer Science Press.

Werbos, P.J. (1974) 'Beyond regression: New tools for prediction and analysis in the behavioral sciences', Unpublished Ph.D. Thesis, Harvard University.

Werbos, P.J. (1988) 'Generalization of backpropagation with application to a recurrent gas market model', *Neural Networks*, 1, 339–56.

Whitley, D. (1989) 'The GENITOR algorithm and selection pressure: Why rank-based allocation of reproductive trials is best', in J.D. Schaffer (ed.) *Proceedings of the Third International Conference on Genetic Algorithms and their Applications (ICGA)*, pp. 116–21, San Mateo, CA: Morgan Kaufmann.

Whitley, D. and C. Bogart (1990) 'The evolution of connectivity: Pruning neural networks using genetic algorithms', *Proceedings of the International Joint Conference on Neural Networks, Washington, DC*, vol. 1, pp. 134–7, Hillsdale, NJ: Lawrence Erlbaum.

Whitley, D. and T. Hanson (1989) 'Optimizing neural networks using faster, more accurate genetic search', in J.D. Schaffer (ed.) *Proceedings of the Third International Conference on Genetic Algorithms and their Applications (ICGA)*, pp. 391–6, San Mateo, CA: Morgan Kaufmann.

Whitley, D. and T. Starkweather (1990) 'Optimizing small neural networks using a distributed genetic algorithm', *Proceedings of the International Joint Conference on Neural Networks, Washington, DC*, vol. 1, pp. 206–9, Hillsdale, NJ: Lawrence Erlbaum.

Wickelgren, W.A. (1981) 'Human learning and memory', *Annual Review of Psychology*, 32, 21–52.

Widrow, B. and M.E. Hoff (1960) 'Adaptive switching circuits', *Institute of Radio Engineers, Western Electronic Show and Convention, Convention Record*, 4, 96–104.

Winograd, T. (1975) 'Frame representations and the declarative–procedural controversy', in D.G. Bobrow and A.M. Collins (eds) *Representation and Understanding: Studies of cognitive science*, pp. 185–210, New York: Academic Press.

Witbrock, M. and M. Zagha (1989) 'An implementation of back-propagation learning on GF11, a large SIMD parallel computer', *Internal Rep.* CMU-CS-89-208, School of Computer Science, Carnegie Mellon University.

Witherspoon, D. and M. Moscovitch (1989) 'Stochastic independence between two implicit memory tasks', *Journal of Experimental Psychology; Learning, Memory, and Cognition*, 15, 22–30.

Wolpert, D.H. (1990a) 'Constructing a generalizer superior to NETtalk via a mathematical theory of generalization', *Neural Networks*, 3, 445–52.

Wolpert, D.H. (1990b) 'The relationship between Occam's Razor and convergent guessing', *Complex Systems*, 4, 319–68.

Wolters, G. (1984) 'Memory: Two systems or one system with many subsystems?', *The Behavioral and Brain Sciences*, 72, 256–7.

Wolters, G. and R.H. Phaf (1990) 'Implicit and explicit memory: Implications for the symbol-manipulation versus connectionism controversy', *Psychological Research*, 52, 137–44.

Zadeh, L.A. (1987) *Fuzzy Sets and Applications: Selected papers*, New York: Wiley.

Zeki, S. (1980) 'The representation of colours in the cerebral cortex', *Nature*, **284**, 412–18.

Zeki, S. and S. Shipp (1988) 'The functional logic of cortical connections', *Nature*, **335**, 311–17.

Name index

235

Subject index

acetylcholine, 54–5
activation–elaboration learning, 10–11, 13, 65–7, 119
activation rule in CALM, 16
Adaptive Resonance Theory, 5, 10, 31, 159, 164, 194
affine transformations, 92, 93, 100
amygdala, 10
amnesia
  anterograde, 64, 81–5
  role of hippocampus, 82, 125
analog hardware, *see* implementation
annealing, 12, *see also* simulated annealing scheme
architectures, 113–14, 135
  and interference, 142–51
ARIMA models, 88
arithmetic, 135, 136
arousal, 10, 11, 13
  process, 24–5, 112–13, 121, 123
  response, 12
  self-induced, 119
  system, 24–5, 112–13, 121, 123
ART, *see* Adaptive Resonance Theory
articulatory loop, 135
artificial subjects, 71, 76, 79–80
attention
  as a functional constraint, 9–11
  arousal function, 10
  selective, 10
  as selective enhancement of activations, 11
attractors, 12
automaton
  cellular, 206–7
  finite-state, 133, 135
  push-down, 133–5

background activation, 20
backpropagation, 4, 5, 6, 7, 106, 128, 135–51, 152, 157, 158, 159, 164, 189, 204
  inverse method, 96
basket cells, 9, 124
batch processing simulations, 187, 201

bidirectional associative memory, 194
biocomputing, 100
blocking of learning, 136
Boltzmann machine, 12, 101, 128
brain-damage, 7
broom balancing, 104
BSP400, 158, 180–90
building block, 103, 104
bus-oriented system, 171

CALM
  discrete version, 119, 185–6
  functioning, 22–33
  physical implementation, 194–7
  status, 119–20
  structure, 15–22
  and supervised learning, 43
CALM Development System, 166, 169, 198, 199–201
CALMLIB, 115, 198, 201–3
CALSOM, 49–55, 119
cart centring, 104
catastrophic interference, *see* interference
categorization
  autonomous, 9
  in CALM, 25–8
chunking, 25
'cloning' of artificial subjects, 76, 79–80
CM–2, *see* Connection Machine
coarticulation, 91
communicating sequential processes, 167
communication
  bottleneck, 171
  deadlock, 169, 173
  overhead, 170, 178
  synchronization, 169, 182, 191, 199, 204
competition, 9, 11, 12, 17, 121
  deadlocks, 12, 26, 186
  indirect, 121–4
  resolving, 25
  structured, 123
  winner-take-all, 93, 126, 159